Camp Cooke
Vandenberg Air Force Base,
1941–1966

ALSO BY JEFFREY E. GEIGER

*German Prisoners of War at Camp Cooke, California:*
*Personal Accounts of 14 Soldiers, 1944–1946*
(McFarland, 1996)

# Camp Cooke and Vandenberg Air Force Base, 1941–1966

*From Armor and Infantry Training to Space and Missile Launches*

JEFFREY E. GEIGER

McFarland & Company, Inc., Publishers
*Jefferson, North Carolina*

LIBRARY OF CONGRESS CATALOGUING-IN-PUBLICATION DATA

Geiger, Jeffrey E., 1952–
    Camp Cooke and Vandenberg Air Force Base, 1941/1966 :
from armor and infantry training to space and missile launches /
Jeffrey E. Geiger.
        p.      cm.
    Includes bibliographical references and index.

    ISBN 978-0-7864-7855-2 (softcover : acid free paper) ∞
    ISBN 978-1-4766-1424-3 (ebook)

    1. Camp Cooke (Calif.)—History.    2. Vandenberg Air Force
Base (Calif.)—History—20th century.    I. Title.
    UA26.C372G45  2014
    355.709794'91—dc23                                        2014002334

BRITISH LIBRARY CATALOGUING DATA ARE AVAILABLE

Front cover photograph: left to right Major Roy A. Peterson, camp
signal officer; Captain Frank A. Partlow, station hospital medical
supply officer; Chief Warrant Officer Clifford F. Breton, chief
clerk, camp finance office; and Master Sergeant William J. Moriarity,
station hospital sergeant major. All four soldiers arrived at Camp
Cooke between October and November 1941. The picture was
taken in October 1944 in front of a billboard on California
Boulevard. U.S. Army/Cooke Clarion

Manufactured in the United States of America

*McFarland & Company, Inc., Publishers*
  *Box 611, Jefferson, North Carolina 28640*
    *www.mcfarlandpub.com*

In memory
of my mother

# TABLE OF CONTENTS

# ACKNOWLEDGMENTS

In writing this book, I owe a debt of gratitude to the following individuals and institutions for their assistance in providing information or photographs that made this book possible. The staff at the National Archives and Records Administration at College Point, Maryland; Dan Sebby, the Curator of the California State Military Museum in Sacramento, California; Bob Peiffer for making available to me his collection of 11th Armored Division images; Myra Manfrina, Karen Paaske, and Readon "Donnie" Gross Silva of the Lompoc Valley Historical Society; the library staffs at the Davidson and Arts Libraries, especially Law Librarian Christopher Granatino, at the University of California at Santa Barbara, for his generous assistance in locating congressional records pertaining to the accident involving Vertie Bea Loggins; Robert A. Lay, Jr., archivist at the Carl Albert Center, University of Oklahoma, for additional information about the Loggins case; Shaun Hayes, assistant archivist at the American Heritage Center, University of Wyoming, for a copy of the Bob Hope script of March 3, 1942; Louise Arnold-Friend of the U.S. Army Military Institute Library at Carlisle, Pennsylvania; Barbara Hall, Jenny Romero, Lea Whittington, and Janet Lorenz of the Margaret Herrick Library in Beverly Hills, California; Kathy Headtke of the Allan Hancock College Library in Santa Maria; Lea Cryor of the Santa Maria Public Library; Ashley Chavez of the Lompoc Public Library; Pam DeTilla of the San Luis Obispo Public Library; Samantha Chin-Wolner of the Museum of Chinese in America in New York City; James A. Tobias of the U.S. Army Center for Military History; and Carol Cepregi of the Congressional Medal of Honor Society.

I am grateful to Cynthia Schur, publisher of the *Lompoc Record*, for permission to use a newspaper photograph in Chapter 5; to William W. Streeter, editor and publisher of the American Banking Association (ABA) *Banking Journal*, and his assistant Andrea Rovira, for providing the image of the Bank of America's office at Camp Cooke, and for granting me permission for its use in Chapter 2; and to Joseph C. Wilder, director of the Southwest Center and editor of the *Journal of the Southwest* (formerly *Arizona and the West*), University of Arizona, for allowing me to reproduce the area map of Camp Cooke/Vandenberg AFB in Chapter 1.

Special thanks to actress Elinor Donohue, who graciously shared with me her recollections of entertaining troops at Camp Cooke.

My thanks also to Nancy Cuddy, the daughter of Camp Cooke commander Colonel Alexander G. Kirby; Pauline Taylor (née Nielson), who, with a group of other young dancers, performed regularly at the Camp Cooke service clubs during World War II; and Philip J. Videon for relaying his experiences with the 747th Amphibious Tank and Tractor Battalion at Camp Cooke from 1950 to 1953.

# PREFACE

This book tells the story of the U.S. Army installation Camp Cooke, beginning with its construction in 1941 and its first nine years, from 1957 to 1966, as Vandenberg Air Force Base. Included in these twenty-five years are two supporting organizations, the U.S. Army Branch Disciplinary Barracks and the Naval Missile Facility Point Arguello. Camp Cooke and Vandenberg AFB are the focus of this history.

Every author starts out hoping to find an abundance of information about his or her subject. When gaps exist in the official records, as with the Camp Cooke army files, one selects the best available information and applies the same rigorous standards of fact-checking before presenting their conclusions to the reader. Thus, research for this book started with Averam Bender's article "From Tanks to Missiles: Camp Cooke/Cooke Air Force Base (California), 1941–1958" and Sergeant Wesley Purkiss' sometimes whimsical unpublished manuscript "A History of Camp Cooke, 1941 to 1946." Purkiss wrote in 1946 using select newspaper articles from the *Cooke Clarion*. Both works provided helpful introductions to the subject.

World War II editions of the *Cook Clarion* are available on microfilm, but the collection is incomplete. More disappointing is that copies of the *Clarion* published during the Korean War were not preserved by any organization. Fortunately, I found a few paper copies of the newspaper for both periods and made good use of them for this book.

My research also led me to spend countless hours reading microfilm copies of the *Lompoc Record*, the *Santa Maria Daily Times*, and the *Santa Barbara News-Press* for articles about Camp Cooke. Trade journals, such as *The Billboard*, were occasionally helpful for information about entertainers who performed at the camp. These publications, some more than others, had a tendency to misspell names. Books about old-time radio stars and vaudeville players became indispensible reference sources not only to avoid repeating the

1

spelling errors, but also for the wonderful background information they provided about the performers. The Margaret Herrick Library has in its collection original and microfilm copies of United Service Organizations files. These were useful in my research about shows performed at Camp Cooke.

The list of Army units that trained at Cooke came from a myriad of sources, notably newspapers, the Purkiss manuscript, published military histories (especially Shelby Stanton's *Order of Battle U.S. Army, World War II*), and various record groups at the National Archives and Records Administration at College Park, Maryland.

For questions I had relating to military training and organizational history, I relied on U.S. Army publications.

One of the more enjoyable aspects of historical research occurs when one surprising discovery leads to a second and then to a third, and so on. This happened after I read a 1944 newspaper account about a tragic accident at Camp Cooke in which an artillery shell exploded, sending shrapnel through a passing Southern Pacific locomotive and injuring two company employees. Though the article had misspelled the name of the more seriously injured victim, Vertie Bea Loggins, I discovered that Congress and the President approved a compensation bill for her. This led me to the congressional reports that detailed the accident and identified Congresswoman Helen Gahagan Douglas as having sponsored the bill. Douglas's congressional papers at the Carl Albert Center, University of Oklahoma, provided additional insight into the case and included correspondence from the victim's attorney, Willedd Andrews. Andrews, who had been the legal counsel for evangelist Aimee Semple McPherson, had a conspicuous history in the Los Angeles courts.

Chapter 1 opens with the federal government sending out soil engineers and surveyors to examine potential sites for Camp Cooke, and subsequently purchasing these properties. Also discussed is camp construction, the arrival of the first station troops, and camp activation in October 1941. Chapter 2 covers the period of December 1941 to June 1946. With the nation on edge following the Japanese attack on Pearl Harbor, and enemy submarines operating off the California coast, this chapter begins with the reaction at Camp Cooke and in the neighboring communities to these events. While much of the chapter examines military training, other topics include the secret radar station south of the camp, war bond rallies, civilian employees, German and Italian prisoners of war at Cooke, and preparations for the proposed invasion of Japan. The chapter ends with the closing of Cooke for the first time, in June 1946. Chapter 3 centers on the interwar years of 1945 to 1950 when the disciplinary barracks, established on the main reservation as an Army prison for military prisoners, also served as temporary caretakers for the vacant camp.

During this time the Army leased out large portions of the camp for grazing. Except for the occasional National Guard and Reserve troops conducting training exercises at the camp, the sound of hoofs and the occasional bleat had replaced gunfire.

The fourth chapter begins with a summary of the Korean War and the reopening of Camp Cooke in August 1950 to train soldiers for overseas combat duty. The large Station Hospital also reopened for the treatment of patients at Cooke and from around the Army, including medical evacuees from Korea. Among other things, the chapter discusses the 1,000 Camp Cooke soldiers who participated in atomic bomb testing at the Nevada bombsite in 1951, and the Army's education program to help overcome illiteracy among its troops.

Chapter 5 describes the civilian entertainment shows that came to Camp Cooke. While the United Service Organizations (USO) and affiliated organizations provided most of the large productions with professional performers, other groups and community talent shows ensured non-stop entertainment for the troops. Large shows featuring notable entertainers of the day are the focus of this chapter. USO clubhouses also opened in the communities around Cooke, giving soldiers additional opportunities to relax and meet new friends. Hollywood movie studios shot scenes for three big productions at Cooke and used hundreds of soldiers as extras.

Chapter 6 examines the closing of Camp Cooke for a final time in March 1953, despite vigorous opposition from the community. The chapter addresses new facility construction at the disciplinary barracks, and its transfer to the U.S. Bureau of Prisons in August 1959.

Chapter 7, "A New Mission," discusses how a deserted Army camp became America's foremost missile and rocket base. It opens with the transfer of North Camp Cooke to the Air Force in June 1957. A year later the Navy received Camp Cooke south of the Santa Ynez River and named it the U.S. Naval Missile Facility Point Arguello. In July 1964 the Navy transferred the entire facility to the Air Force, adding almost 20,000 acres to the base. The purchase of Sudden Ranch in March 1966 increased South Vandenberg in size by approximately 15,000 acres. This brought Vandenberg to its current size of 99,099 acres.

Among the notable missile activities discussed in Chapter 7 is the first launch from Vandenberg AFB in December 1958, the first space reconnaissance satellite, and America's first nuclear missile placed on alert. By the end of December 1966, some 653 vehicles had been launched from Vandenberg. In that year alone, 123 launches were conducted, including the first two pairs of simultaneous or dual Minuteman I intercontinental ballistic missiles, and the first heavy-lift Titan IIIB-Agena D space booster. Along with missile successes came missile accidents and the loss of an airman. Vandenberg's most ambitious program of the 1960s was the ill-fated Manned Orbiting Laboratory that was supposed to be a manned space reconnaissance program. In 1962 the Cuban Missile Crisis brought Vandenberg AFB to a high state of combat readiness, with missiles prepared to launch. Among the most important visitors to Vandenberg AFB were President John F. Kennedy, Vice President Lyndon Johnson, Attorney General Robert F. Kennedy, and Russian Premier Nikita S. Khrushchev, who sped through the base on a locomotive. Since my objective was to show Vandenberg's important role in national defense and its rapid expansion in size and diverse missions during its early years, this chapter offers representative samples of the most salient events from that period.

Some of the documents consulted for Chapter 7 were originally classified secret by the Air Force but were declassified before I consulted them. In instances where the document is still classified, I used only the unclassified information for this book.

## Abbreviations and Acronyms

| | |
|---|---|
| **A1C** | Airman First Class |
| **A2C** | Airman Second Class |
| **AFF** | Army Field Forces |
| **AGF** | Army Ground Forces |
| **ARDC** | Air Research and Development Command |
| **ARPA** | Advanced Research Projects Agency |
| **ASU** | Area Service Unit |
| **ACTH** | Adrenocorticotropic hormone |
| **AWOL** | Absent Without Official Leave |
| **AWVS** | American Women's Voluntary Service |
| **Capt.** | Captain |
| **CASC** | Corps Area Service Command |
| **Col.** | Colonel |
| **Cpl.** | Corporal |
| **CSI** | Camp Shows, Inc. |
| **DEFCON** | Defense Condition |
| **FBI** | Federal Bureau of Investigation |
| **Gen.** | General |
| **GI** | Government Issue |
| **HCC** | Hollywood Coordinating Committee |
| **H.R.** | House Resolution (Congressional) |

| | |
|---|---|
| **HVC** | Hollywood Victory Committee |
| **ICBM** | Intercontinental Ballistic Missile |
| **IRBM** | Intermediate Range Ballistic Missile |
| **ISU** | Italian Service Unit |
| **KH** | KEYHOLE |
| **LCC** | Launch Control Center |
| **LST** | Landing Ship Tank |
| **Lt.** | Lieutenant |
| **Lt. Col.** | Lieutenant Colonel |
| **LVT** | Landing Vehicle Tracked |
| **Maj.** | Major |
| **Maj. Gen.** | Major General |
| **MFCO** | Missile Flight Control Officer |
| **MOL** | Manned Orbiting Laboratory |
| **MP** | Military Police |
| **MSgt.** | Master Sergeant |
| **NASA** | National Aeronautics and Space Administration |
| **NBC** | National Broadcasting Company |
| **NCO** | Noncommissioned Officer |
| **OSTF** | Operational System Test Facility |
| **OWI** | Office of War Information |
| **Pfc.** | Private First Class |
| **PG&E** | Pacific Gas and Electric |
| **PMR** | Pacific Missile Range |
| **POW** | Prisoner of War |
| **PRIME** | Precision Recovery Including Maneuvering Reentry |
| **Pvt.** | Private |
| **RAF** | Royal Air Force |
| **RoK** | Republic of Korean |
| **RP-1** | Rocket Propellant 1/Refined Petroleum 1 |
| **SAC** | Strategic Air Command |
| **SCU** | Service Command Unit |
| **Sgt.** | Sergeant |
| **SLTF** | Silo Launch Test Facility |
| **SLV** | Standardized Launch Vehicle/Space Launch Vehicle |
| **SSgt. (S/Sgt.)** | Staff Sergeant |
| **SWHO** | History Office, 30th Space Wing, Vandenberg AFB, California |
| **TSgt.** | Technical Sergeant |
| **U.K.** | United Kingdom |

| U.N. | United Nations |
|------|----------------|
| USAF | United States Air Force |
| USAFI | United States Armed Forces Institute |
| USDB | U.S. Disciplinary Barracks |
| USO | United Service Organizations |
| WAC | Women's Army Corps |
| WDD | Western Development Division |
| YMCA | Young Men's Christian Association |
| YWCA | Young Women's Christian Association |

# INTRODUCTION

Between September 1939 and early 1941, Nazi Germany had conquered most of Europe and was continuing to expand its reign of death and destruction. In the Far East, Japan was occupying China and French Indochina, and was preparing to strike at the Pacific Islands. Concerned about the growing threat to its security, but unprepared and reluctant to enter the conflict, the United States was now rapidly strengthening its military defenses. On September 16, 1940, Congress enacted the first peacetime draft in the United States that brought large numbers of men into the armed forces. To accommodate the increased size of the military, the War Department began building hundreds of new Army and Navy training centers around the country. One of these installations was Camp Cooke on the Central Coast of California. Today it is Vandenberg Air Force Base and is the nation's only facility for launching satellites into near polar orbit and test firing Intercontinental Ballistic Missiles to targets in the Pacific Ocean.[1]

Plans for establishing Camp Cooke began in March 1941 when the U.S. War Department authorized the Army to find a site in California to build the first West Coast training installation for armored divisions. A few weeks later the Army reported that it had selected a rural site 150 miles northwest of Los Angeles in Santa Barbara County between the communities of Lompoc and Santa Maria. In May 1941 it sent a surveyor team into the area to map the terrain and prepare for construction. Over the next few months the Army began purchasing lands, and in mid–September 1941 broke ground for the new camp. Additional properties acquired in early 1942 ultimately increased the size of the camp to between 92,000 and 94,000 acres.[2]

Though construction was still months away from completion, the Army activated the camp on October 5, 1941, and named it Camp Cooke in honor of Maj. Gen. Phillip St. George Cooke, a cavalry officer who served in the Mexican War, Indian Wars, and Civil War. A second construction phase, which

7

started in August 1942 and ended in February 1943, brought the final cost of building the camp to more than $32 million.[3]

Camp Cooke bestrode parts of several nineteenth century Mexican land grants. At more than 40,000 acres, the sprawling Rancho de Jesus Maria was the largest of these grants, and almost all of it was enclosed within the northern part of the camp. After the Mexican War (1846–1848), the Territory of California transferred to the United States. A few years later a savvy businessman named Lewis T. Burton purchased Jesus Maria and used it to develop a lucrative ranching business. Much of the property consisted of a mesa that was later eponymously named Burton Mesa. By the turn of the century, Rancho de Jesus Maria had again changed hands. In 1906, Edwin Jessop Marshall bought the property and developed it into Marshallia Ranch in the years leading up to World War II. Marshallia would later serve as the residence of Army leadership at Camp Cooke.[4]

The 5th Armored Division rolled into Camp Cooke in February 1942. With the exception of a four-month period when it moved to the California desert for advanced maneuvers, the division trained at Cooke continuously until March 1943 when it deployed to Europe. It was followed by the 6th Armored Division (March 1943–January 1944), the 11th Armored Division (February 1944–September 1944), the 86th Infantry Division (September 1944–November 1944, including seven weeks at Camp San Luis Obispo), the 97th Infantry Division (September 1944–February 1945), the 13th Armored Division (July 1945–November 1945), and the 20th Armored Division (August 1945–March 1946). Also at Cooke was the 2nd Filipino Infantry Regiment (later redesignated as a battalion), and numerous other specialties, including antiaircraft artillery, combat engineer, ordnance, and hospital units.[5]

In order to prepare its soldiers for the physical and emotional demands they would likely encounter on the battlefield, the Army's training program at Cooke and at other installations was a conglomeration of many elements. It included instructing each man in weapons proficiency and personal survival, strengthening their physical endurance, and instilling a sense of personal confidence and unit adhesiveness as a fighting organization. The Army sought to motivate and convince each man that he was fighting for a worthy cause. In this regard, the Japanese attack on Pearl Harbor had already sparked a national outrage and thirst for retribution. The Army tapped into this sentiment in lectures and films that exposed the aggression and atrocities of Japan and Nazi Germany. During the Korean War, the motivating factor was the fear of Communist domination.

An important subset to all troop motivation is maintaining high morale among the ranks. In this area, the Army made good use of entertainment and recreation, and encouraged religious participation by its troops. Collectively, these elements became the key factors in building combat troop readiness.

A weekly camp newspaper, the *Cooke Clarion*, helped to promote the values and unit cohesiveness sought by the Army. The paper provided its readers with a steady stream of information about the comings and goings of organizations on the post, social and chapel activities, commentary, war news, sports, and special events.

Sports became a popular pastime at the camp, with many units forming their own baseball and basketball teams that competed on the installation and against teams at other military bases in California. Amateur boxing was another popular activity practiced among soldiers at indoor and outdoor rings. On two occasions while touring military bases for the Army's Special Services Branch, world heavyweight boxing champion Sgt. Joe Louis staged exhibitions at Cooke, to cheering crowds.[6]

To host entertainment shows and other festive events at Cooke, the Army built a Sports Arena (sometimes called a Field House), service clubs, and theaters. The two service clubs offered nightly venues that included floor shows, dances, sing song sessions, ping pong events, hobby nights, movie nights, and game nights. Each club also housed a library of several thousand books. Elaborate stage shows that attracted large audiences were held at the Sports Arena.

Entertainers who performed at Cooke read like a who's who list of Hollywood luminaries and popular celebrities. Countless other entertainers from radio shows and stage acts—including dance choruses of pretty girls, comedians, musicians, singers, jugglers, and acrobats, many of their names long ago faded into obscurity—thrilled and delighted the thousands of GIs that passed through Camp Cooke. For many GIs, particularly those from rural parts of America, these shows gave them their first exposure to legitimate stage performances.

Most of the traveling extravaganza shows presented at Cooke and at other Army and Navy facilities in the United States and abroad during World War II were sponsored by the United Service Organizations–Camp Shows, Inc. The USO and its affiliate Camp Shows, Inc., were officially designated by the War Department to provide entertainment for the armed forces. At its core, the USO consisted of six civilian relief agencies that coordinated their war efforts and resources to become the GI's "home away from home." The six agencies were the Salvation Army, the Young Men's Christian Association, the National Catholic Community Service, the National Jewish Welfare Board, the Young Women's Christian Association, and the National Travelers Aid Association.[7]

Camp Shows, Inc., which grew out of a separate earlier organization, became an official affiliate of the USO on October 30, 1941. Headed by leaders of the entertainment industry from stage, screen, and radio, CSI was financed by the USO to arrange and send out entertainment requested by the Special Services Branches of the military.[8]

Another associate organization of the USO serving on the West Coast was the Hollywood Victory Committee. Composed of actors, writers, and producers, it was formed three days after the attack on Pearl Harbor. The HVC recruited a voluntary talent pool of performers that made up CSI troupe tours. Sometimes HVC sent celebrities directly to military installations. It also arranged bond drives, hospital tours, patriotic radio broadcasts, and war relief campaigns. CSI and HVC closed at the end of World War II. They were revived in 1951—the latter organization had become the Hollywood Coordinating Committee—and served the military during the Korean War. The HCC closed permanently in 1953, followed by CSI in 1957.[9]

Also sponsoring variety shows at Camp Cooke and at other military installations were large companies such as Shell Oil and R. J. Reynolds Tobacco, and well-connected individuals including Larry Crosby, the brother of Bing Crosby.[10]

Hollywood came to Camp Cooke to film scenes for three major movies, beginning in February 1943 with Irving Berlin's *This Is the Army*, followed by *In the Meantime, Darling*, in December 1943 and *Counter-Attack* in November 1944. In each film dozens of soldiers filled in as extras.[11]

Equally enjoyed by thousands of soldiers at Cooke were the USO service clubs in the neighboring communities of Lompoc, Surf, and Santa Maria. During World War II, Lompoc boasted the presence of two USO clubs. A third club at Surf on the outskirts of Lompoc was one of the few USO facilities permitted to operate on a military installation. Always a source of comfort, these clubs became many things to many people: a place for entertainment; recreation; dancing and meeting people; to see movies or find religious counsel; a quiet place to talk or write letters; assistance with child care or housing; and, of course, the place to go for free coffee and doughnuts. Many of the talented performers at these clubs came from the neighboring communities.

"Weekend house parties" were another fun-filled opportunity for soldiers and young ladies to meet. These events were usually arranged by unit special services officers and USO clubs in Los Angeles. Anywhere from 75 to more than 150 girls would be bussed to Camp Cooke, where they enjoyed dinner, dancing and tours of the camp provided by homesick soldiers. For one event the 212th Armored Field Artillery used the occasion to put on an amateur talent show for their guests.[12]

Community organizations, many with nationwide affiliations like the

American Women's Voluntary Service, the American Red Cross, the Gray Ladies Society, and the American Legion, provided a largess of activities and assistance to soldiers. The AWVS sponsored dances at USO clubhouses, American Legion posts, and at the Lompoc Veterans Memorial Building. They also trained junior hostesses in the proper etiquette for speaking and dancing with soldiers, and serving refreshments. The Gray Ladies frequently worked with the Red Cross in aiding GIs, especially the hospitalized.

During World War II the Hancock Music Ensemble, with roots in Southern California and Santa Maria, gave concerts at Camp Cooke. Another contributor was Marjorie Hall. She owned a dance school in Santa Maria and frequently appeared with her students in dance recitals at the two service clubs and at the Station Hospital. She also taught dance at Service Club No. 2.[13]

As entertainment chairperson of the American Red Cross Camp and Hospital Service Council, Irene Mesirow and later Emma Stockton brought together talented residents from Santa Maria, Casmalia, and Guadalupe to perform in musical variety shows that played to patients and staff at the Station Hospital. Other community volunteers helped by donating recreational equipment to USO facilities, and furnishing the many dayrooms and two guest houses at Cooke.[14]

Probably no other individual in Santa Barbara County did more for service members at Camp Cooke than Faye Porter. As entertainment director of the local chapter of the Works Projects Administration War Service Unit, Porter brought countless talent shows and activities from around the county to the service clubs at Camp Cooke. Her appointment in August 1943 as director of the USO club at Surf gave Porter an additional outlet for bringing entertainment to service members. During the Korean War she resumed relief work for military members at Cooke as an employee of the Lompoc Community and Servicemen's Center.[15]

To the GIs who knew Mrs. Porter, she was "Mom." Born in Minnesota in June 1883, she came to Santa Barbara in 1913 after graduating from Lawrence Conservatory of Music in Kansas. Porter eventually settled in Lompoc, where she died on April 4, 1974, survived by a son and two granddaughters.[16]

Beginning in 1944, German and Italian prisoners of war were interned at Camp Cooke. After the surrender of Italy in September 1943, many Italian prisoners volunteered to work for the U.S. Army. They were organized into Italian Service Units and shipped to Army camps throughout the United States. The Germans and Italians were kept separate from each other in accordance with the Geneva Convention. Since many Italians were now working for the Americans, the especially hardcore Nazis viewed them as turncoats.[17]

By June 1943, the number of American soldiers at Cooke had climbed to 35,781. Thereafter the population slowly declined to less than two hundred soldiers by the time the camp closed for the first time on June 1, 1946.[18]

In December 1946 the Army established a maximum security branch U.S. Disciplinary Barracks on post property to confine military prisoners, mostly Army, serving court martial sentences. The disciplinary barracks was also responsible for camp security and caretaker services during the two times that Camp Cooke was closed, once after World War II and again after the Korean War.[19]

Six weeks after the outbreak of the Korean War, Camp Cooke was reactivated, on August 7, 1950. Two National Guard infantry divisions, the 40th from California, and the 44th from Illinois, were the largest organizations stationed at Camp Cooke during the war. The 40th was inducted into federal service in August 1950 and moved to Cooke a month later for active duty training. In March 1951 it departed for Japan and later served on the front line in Korea. The 44th Infantry Division was also federalized and arrived at Cooke in February 1952. A few months later it was reclassified from a combat readiness mission to replacement training—training soldiers for other units serving overseas. In November 1952 the division moved to Fort Lewis, Washington.[20]

Dozens of smaller organizations, mostly detached reserve units, were at Cooke for several weeks and occasionally up to two years, including the 747th Amphibious Tank and Tractor Battalion from Florida, and one of the last remaining segregated black units in the Army, the 466th Antiaircraft Artillery Battalion from Virginia.[21]

The hospital's annual medical activity reports for Camp Cooke indicated the military population for 1951 and 1952 averaged 10,357 and 13,871, respectively. The month with the highest number of soldiers on the instillation appeared to be February 1951, peaking at 20,569.[22]

Supplementing USO entertainment shows at Cooke were an assortment of amateur talent from the local community, and stage performances by soldiers for fellow khaki-clad service members. They sang, danced, played instruments, told jokes, and acted in comedy skits. Sometimes soldier performances were held as talent contests, with prizes awarded to the best players.[23]

With the war in Korea nearing an uneasy truce, the Army closed Camp Cooke for a second time on March 31, 1953. Following a policy established at Cooke after World War II, the Army again leased large sections of the camp for grazing and agriculture.

In June 1957 the Army transferred the northern half of the camp, excluding the area around the disciplinary barracks, to the Air Force for use as a missile base. Initially called Cooke Air Force Base, it was renamed Vandenberg

**Groundbreaking and dedication day at Camp Cooke. September 14, 1941. U.S. Army/ author's collection.**

of supervisors; Samuel J. Stanwood and Monroe Rutherford, Santa Barbara County supervisors; Walter McIntosh and Francis P. O'Reilly, Santa Barbara City councilmen; Charles A. Storke, associate publisher of the *Santa Barbara News-Press*; Edward H. Stamm, president of the Santa Barbara Chamber of Commerce; and Robert Fisher of the Southern Pacific railroad. Also attending the program was Father Augustine Hobrecht, president of the St. Augustine's Seminary in Santa Barbara.[5]

The program ended with Lt. Col. Steele shouting out, "Give 'er the gun," the signal to heavy equipment operators to fire up huge carryalls and begin gouging yards of topsoil from the Burton Mesa in a symbolic groundbreaking gesture. A flyover by three Army PT-13 Stearman aircraft from Hancock College of Aeronautics in Santa Maria capped the festivities.[6]

Actual site work began the next day on September 15, 1941, with fleets of earth moving equipment clearing away dense brush and leveling the ground for roads, utilities, and buildings. To overcome the problem of building on sandy soil, thousands of tons of shale were carved out of the countryside and mixed into the sand at construction sites to create firm foundations. Another

*Top:* A portion of the multiple carloads of materials that arrived each day at the railroad siding at Tangair for the building of Camp Cooke. Workmen shown in the picture stack the lumber in piles ready for carriers to haul it to the project. September 26, 1941. U.S. Army/author's collection. *Bottom:* Pictured above is part of the approximately fourteen miles of railroad track that was built from the Tangair siding into Camp Cooke. Its completion in February 1942 came in time to handle the first large troop movements into the camp. October 27, 1941. U.S. Army/author's collection.

early difficulty was transportation. Materials were brought into the camp by truck and rail, and then stockpiled at Tangair. Until the railroad spur from Tangair to the building areas in the camp was completed, convoys of trucks carried the materials to the worksites, but at times they were barely keeping up with crews in the field. Sporadic wind gusts created swirling clouds of sand that slowed work on many parts of the project. Winter rainstorms added to the difficulties.[7]

The lack of water and the absence of electrical power were the two largest obstacles that hampered construction during the first three weeks of the project. At first, tanker trucks were used to haul water into the camp from Lompoc, but their limited capacity held down concrete production to a fraction of the 1,000 yards needed daily to keep ahead of carpenter crews. Occasionally it became necessary for concrete workers to pull an additional night shift to speed up work. The situation improved by early October after workers installed a temporary pipeline from the recently completed Number 4 water well near the Santa Ynez River to a storage tank in the camp. Trucks could now quickly refill and bring water to the worksites for concrete mixing. Meanwhile, crews had extended the pipeline to a large concrete mixing plant that, when completed in mid–October and powered by electricity, began producing the required quota.[8]

Pacific Gas and Electric (PG&E) brought electrical power to the main entrance of the camp in early October. The construction subcontractor in charge of power service inside the camp extended the power throughout the installation, and connected with the concrete mixing plant, the sawmill plant, and other machinery. Power greatly aided in speeding up the work and the hiring of additional work crews.[9]

In mid–October 1941, about 2,300 workers were on the job constructing the camp. The workforce grew to more than 5,000 employees before the project ended in February 1942. With a workforce almost double the population of Lompoc, it wasn't surprising that a housing shortage forced many of the workers to live in tents and trailers close to their worksites. A large "Tent City," with a commissary and dining area near the main gate, also housed workers.[10]

On September 24, 1941, the War Department announced that it selected the name Camp Cooke, in honor of Maj. Gen. Philip St. George Cooke, for the Santa Maria–Lompoc army camp, currently under construction. Born in Virginia in 1809, Cooke was a cavalry officer and had served most of his career on the frontier until his retirement from the Army in 1873. During the Civil War he remained loyal to the Union Army and painfully watched as almost his entire family joined the Southern cause.[11]

*Top:* The commissary at "Tent City" at the front of the camp. The sign on the left indicates that meal tickets are available for "strictly cash in advance." September 23, 1941. U.S. Army/author's collection. *Bottom:* Army Signal Corps linemen from the 255th Signal Construction Company at Fort Ord installing telephone poles at Camp Cooke. October 30, 1941. U.S. Army/author's collection.

**The first telephone switchboard at Camp Cooke was operated by enlisted men from Company B, 54th Signal Corps Battalion, Fort Ord. October 31, 1941. U.S. Army/ author's collection.**

Ironically, Camp Cooke was separated into northern and southern sections by the Santa Ynez River. On the south side, close to where the estuary meets the Pacific Ocean, is Ocean Beach Park, sometimes called Ocean Park. Just below the park sat the tiny community of Surf with its unadorned railroad station. East of the park was the settlement of Baroda. In October 1941, construction began on a 29,000-foot-long salt water barrier bridge at Baroda. The structure not only connected the two sections of Camp Cooke, it also acted as a sea water retaining wall that allowed water to flow downstream while keeping contamination from flowing upstream to the five fresh-water wells that were being built on the north side of the shoreline. The bridge consisted of wood pilings, concrete, and earthen fill. The center section included concrete flumes and spillways, with one or two fish ladders to allow the annual steelhead trout run. By early 1942, the wood planking of the bridge, the last of the structural chores, was complete. The northern end of the bridge connected to Barrier Road, which was an extension of 35th Street. The southern portion near Baroda met Highway 150, today known as 246.[12]

By late 1943, frequent tank and truck crossings on the bridge had exco-

An Air Force convoy transporting a Titan missile south across the salt water barrier bridge at Vandenberg AFB. The Air Force renamed the span Surf Bridge. May 22, 1968. U.S. Air Force.

riated much of the planking. Army engineers resurfaced the structure using half-inch boiler plates. The bridge remained in use until January 1969 when a massive rain-swollen flood in the Lompoc Valley destroyed the structure. It was never rebuilt.[13]

On October 5, 1941, the Army's Ninth Corps Area headquarters at the Presidio in San Francisco, California,[14] issued General Orders No. 87 directing the activation of Camp Cooke and Corps Area Service Command (CASC) 1908 as the post headquarters unit to operate the camp. A few days later, Capt. Roswald F. Smith and eleven enlisted men arrived at Cooke as the first permanent contingent of troops. They probably came from the headquarters service detachment at the Presidio in San Francisco. With barracks still under construction at Cooke, they set up a tent camp several hundred yards west of Wyoming and Ocean View avenues, and for the next two months resided under a canopy of blue-gum eucalyptus trees. They aptly called their accommodations "Blue Gum Terrace."[15]

The men cooked their meals over an open fire, and for drinking and wash-

*Top:* Workers constructing the southern extension of Washington Avenue known as Pine Canyon Road that connected the camp with the nearby town of Lompoc. November 3, 1941. U.S. Army/author's collection. *Bottom:* California Boulevard and 26th Street, with Theater No. 1 on the left under construction. December 15, 1941. U.S. Army/author's collection.

ing relied on water trucked in daily from Lompoc. Passes were issued to the neighboring towns for hot baths. Twice each week, trucks were dispatched to Camp San Luis Obispo for commissary and laundry services.[16]

By the first week of October 1941, the Associated Telephone Company

*Top:* Civilian contractors framing a building. November 5, 1941. *Bottom:* Installing water and sewage lines. February 17, 1942. U.S. Army/author's collection.

Pictured above is Camp Cooke in late October or early November 1941 during phase 1 construction, which stretched from the middle of the camp to the far end at 35th Street. When completed in early 1942, much of this area was occupied by the 5th Armored Division, the first Army Ground Forces troops to arrive at the camp. In the foreground are "Tent City" and a few temporary buildings that included a commissary, a dining area, and housing for civilian construction workers. To the right is California Boulevard, the main roadway through the camp. The large rectangular development at the upper right is the hospital area. Toward the end of this construction phase, preliminary roads were laid out and utility lines installed at the front half of the camp to the right of California Boulevard. This work was in preparation for phase 2 facility construction that began in August 1942 under a different set of contracts. U.S. Army/author's collection.

had established telephone service to the main gate at Camp Cooke and was continuing to install additional trunk lines. The task of bringing those outlets into the camp and connecting it with the outside world began in late September when 1st Lt. William Jennings and a group of enlisted men from the Army's 54th Signal Corps Battalion at Fort Ord, in northern California, arrived at the camp to begin stringing lines to the various parts of the construction area. They brought with them a temporary field communications trailer and a telephone switchboard. By the middle of October, additional soldiers from the 255th Signal Construction Company were installing permanent telephone poles and stringing cables. This latter detachment of eighty-nine men included kitchen, supply, and administrative personnel led by 1st Lt. Delos W. Sellon, 2nd Lt. Stanley Doran, and MSgt. William Angel.[17]

The image directly above shows the construction of roadways and facilities in the mid-section of the camp as of November 18, 1941. The industrial area under construction in the foreground consisted mostly of warehouses, storage and maintenance shops, and a large loop of railroad track used to bring supplies into the camp from the Southern Pacific Railroad siding at Tangier. U.S. Army/author's collection.

On October 20, 1941, Lt. Roy A. Peterson arrived at Cooke with a signal construction company that became part of the camp's permanent contingent as the Signal Office for the Post Telephone System. The system gradually matured and on March 11, 1942, was cut over, giving Cooke direct outside service through the local switchboard. Construction continued on the outside plant until 45 miles of trunk cable and 65 miles of aerial wire had been put into operation. Eventually, close to 800 lines were in use, with a staff of more than twenty operators on hand to handle the calls. The Associated Telephone Company took over the maintenance and operation of the system on April 1, 1942, and continued to operate the switchboard until March 31, 1946, at which time the operations portion of the contract was terminated. The company continued to maintain the local telephone plant probably until the camp closed two months later.[18]

Camp Cooke's first commanding officer was Lt. Col. John B. Madden. He arrived at the camp on October 15, 1941, and, finding no buildings ready

Peering across Camp Cooke from its southwestern edge at 35th Street. The four parallel roads, from left to right, are California Boulevard, Arizona Avenue, Nevada Avenue, and New Mexico Avenue. A portion of the railroad tracks is visible along the rim south of 35th Street and to the right of New Mexico. November 18, 1941. U.S. Army/author's collection.

for occupancy, moved his staff twenty-five miles north to Santa Maria. There, Madden established CASC 1908 headquarters in the Chamber of Commerce building at 110 South Lincoln near Church Street. In late November they moved into the still incomplete post headquarters building on California Boulevard near Utah Avenue.[19]

November 20, 1941, was a crisp Thursday morning when the entire garrison of eleven officers and thirty enlisted men stood at attention during the first flag raising ceremony outside the post headquarters. The 35-foot-high pole flying the American flag was hewed from a eucalyptus tree. By the end of the year, the men at Blue Gum Terrace traded their alfresco home for barracks. The detachment numbered about eighty enlisted men.[20]

As 1942 dawned, Camp Cooke received a new post commander, Lt. Col. Carle H. Belt. He assumed the responsibility on January 7, and was promoted to full colonel three months later. A veteran of more than twenty years in the military, Col. Belt was recalled to Army service in 1941.[21]

During Col. Belt's tenure as camp commander, Cooke underwent a major

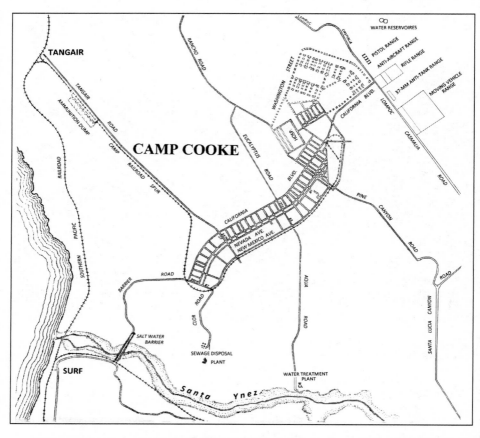

This reproduction of an early 1942 map of Camp Cooke shows the camp still under construction. Several of the secondary roads in the main cantonment area are missing from this image. The dotted lines between Washington Avenue and California Boulevard at the front of the installation are barracks and other facilities that were planned for phase 2 construction beginning in August 1942. Washington Street was renamed Ocean View Avenue, probably during the new construction. U.S. Air Force.

expansion of its facilities. It began modestly in April 1942 when the federal government allocated approximately $399,000 to the Works Projects Administration agency to construct new roads, sidewalks, and firebreaks, and to upgrade select buildings for use as officers' quarters. Three months into the project, the government abruptly terminated the work, on July 29, and awarded essentially the same roadway work to MacDonald & Kahn on a ninety-day contract for $301,187.83. On that same day, the Army's District Engineer Office in Los Angeles announced that a second major construction project at Cooke would begin shortly and nearly double the housing capacity of the camp.[22]

Sure enough, in August 1942 a new construction project started that encompassed the section between Colorado Avenue and the main gate. During the first construction project, water and power systems were extended into this area, but the actual building portion was deferred. Unlike the first construction project, which was issued as a single contract to two companies in a joint venture, the new project was divided among multiple companies, with each firm receiving a separate contract for specific work. The expansion project included barracks, mess halls, chapels, a third and fourth theater, a second service club, a second guest house, and additional motor pools. The Theater of Operations Barracks constructed in the new section were flimsier than the two-story Mobilization Barracks erected during the first construction phase, leading soldiers to disparagingly nickname this part of the camp "Splinterville." Completed in February 1943 at an estimated cost of $15 million, it brought the total cost for the two construction phases to more than $32 million.[23]

# 2

## THE CAMP IN WORLD WAR II

On December 7, 1941, Japan attacked the American naval base at Pearl Harbor in the Hawaiian Islands, immediately bringing the United States into World War II. Days later, the Axis powers of Germany and Italy came to the support of Japan and declared war on the United States. The war changed the situation at Camp Cooke and in communities across the country. At Cooke, armed military and civilian guards set up barriers around the camp and patrolled the reservation, with orders to shoot anyone who failed to obey halt signals. In Lompoc, citizen volunteers organized a civilian defense council. Working with military officials and civilian police, they established air raid observation posts, with residents taking turns watching for enemy ships and aircraft. Armed civilian guards were stationed at the city's water supply; air raid wardens were assigned to each block in the community; and a blackout law was initiated, with the first test on December 10. The blackout order issued by the Army extended throughout Southern California.[1]

Japanese-Americans and Japanese aliens immediately fell under suspicion as potential enemy agents. In the Lompoc and Santa Maria valleys the Japanese were predominantly farmers growing vegetables on leased land. The cordial relations that existed between Caucasians and Japanese in these communities turned in many instances into fear, distrust, and even anger against the local Japanese after the attack on Pearl Harbor. Within hours following the attack, the Federal Bureau of Investigation swooped down on Santa Maria, arrested fourteen of the city's most prominent Japanese citizens, and carted them off to Los Angeles for interrogation. Five days later, Japanese aliens in Lompoc were required to register at the police station. On February 18, 1942, the FBI arrested some 250 Japanese men in Guadalupe, near Santa Maria, and at least 23 men in Lompoc the following day. Arrests of Japanese residents were occurring in other parts of the country and as far away as New York. Japanese submarines operating off the West Coast of the United States were making the situation worse. In the spring of 1942, Lompoc's Japanese community of about one-hundred families, and thousands of other Japanese on the West Coast, were forced by the federal government to leave their homes and sent to internment camps away from the coast for the duration of the war.[2]

On the warfront in 1942, the United States and its Allies were mostly on the defensive, but were occasionally winning battles. In the Pacific, the Japanese entered the Philippine capital of Manila in January, and in May had overwhelmed American and Filipino defenders on the peninsula of Bataan and the Island of Corregidor. The one bright spot for the Americans came a month later at the Battle of Midway where the United State achieved a decisive naval victory. In August, the United States landed Marines on Guadalcanal, beginning a brutal six-month battle of jungle warfare. It ended with the Japanese evacuation of the island in February 1943. In North Africa, the fighting between Field Marshal Erwin Rommel's Afrika Korps and British troops seesawed back and forth during most of 1942. When American troops arrived in North Africa in November to aid the British, the German-Italian forces were squeezed back to Bizerte, Tunisia, where they surrendered to Allied forces in May 1943.[3]

Back in the United States, in October 1941 the 5th Armored "Victory" Division was activated at Fort Knox, Kentucky, under Maj. Gen. Jack W. Heard. Five months later, on February 16, 1942, the first elements of the 5th detrained at Camp Cooke and went by truck convoy to rows of freshly constructed two-story barracks. Nearby, on California Boulevard between 15th and 18th Streets in building 6007, stood Division Headquarters, flanked on both sides by officers' quarters. Directly behind headquarters was the Officers' Club, and to the front across the parade grounds was Service Club No. 1 with its columned porch.[4]

The armored division spent almost ten months training at Cooke and three additional months, between August 8 and November 19, 1942, participating in grueling maneuvers at the Desert Training Center east of Riverside in Southern California. On March 17, 1943, it departed Cooke for camps on the East Coast and then on to England in February 1944. Six months later the division landed in France.[5]

In addition to the 5th Armored Division, at least fourteen other detached units from battalion level and below arrived at Cooke in 1942. With so many soldiers all looking for something to do, some place to go, and the opportunity to talk with young ladies, Cooke's Special Services officer, Maj. Virgil E. Reames, led the effort to establish a camp newspaper that would provide its readers with information about entertainment shows, unit activities, sports

Troops of the 5th Armored Division arriving at Camp Cooke. February 16, 1942. U.S. Army/author's collection.

The Division Headquarters, located in building 6007 on California Boulevard between 15th and 18th Streets, housed the commanding general of the 5th Armored Division and the leadership of subsequent divisions that trained at Camp Cooke during World War II and the Korean War. Shown above is the building shortly after its completion in 1942 with its fresh coat of white paint before being repainted Army drab brown. U.S. Army/author's collection.

news, special events, and other happenings around the installation. Under the quizzical masthead *What's My Name?* the first edition rolled off the press on March 13, 1942, announcing the start of a contest to select a name for the paper. Five days later, Maj. Alfred Walter and MSgt. Benjamin Bosley submitted the winning suggestion: the *Camp Cooke Clarion*. Both men shared first and second prize money equally, totaling $25. Third prize went to the camp Finance Office for suggesting the distinctive red "V" in the paper's masthead. The paper was published for the Army every Friday by the *Santa Maria Daily Times*.[6]

In 1943 the newspaper changed names twice, first becoming *The Clarion* and finally, two months later on May 21, the *Cooke Clarion*. Accompanying the latter change was a new masthead, designed by Pvt. Pagsilayan (Pagsilang) Rey Isip of the 2nd Filipino Infantry Regiment. It depicted an allied soldier gleefully chasing Adolf Hitler, Benito Mussolini, and Tojo at the point of a bayonet. The centerpiece of massed flags represented the unity of nations. Pagsilayan was a successful artist for science fiction and fantasy magazines, and for poster art in the 1930s and after the war.[7]

Another early addition to the camp were guest houses. They provided clean, comfortable lodgings at very affordable rates to families visiting their loved ones stationed at Cooke. The Army understood that family reunions also helped to improve morale and order among its ranks, and quite possibly lowered the incidence of homesick soldiers going AWOL. The first of two guest houses at Cooke opened on April 5, 1942, in building 6105, a two-story, 7,840 square foot structure at the corner of Arizona Avenue and 15th Street. The second guest house, built during the second phase of camp construction, opened a year later, on April 4, 1943, near the main gate. The new building was probably smaller and had 25 or more bedrooms. Both guest houses were fully furnished and decorated by the generous donations of community residents and civic organizations, especially women's benevolent groups. They included the Scandinavian committee, students from Santa Barbara State College, the American Legion Auxiliary, the Canadian Legion Auxiliary, and the Santa Barbara chapter of the National Women's Relief Corps. The room rate was 75 cents per day for adults and 50 cents per day for children under the age of twelve sharing rooms with their parents.[8]

Guest houses operated under a set of rules similar to civilian hotels, but with two major differences. Guests were limited to a three-day visit that could be extended under certain conditions. Officers and enlisted men were prohibited from entering any of the rooms except for the reception area.[9]

The grand opening of the first guest house at Cooke coincided with the observance of Army Day held on April 6, 1942. The 5th Armored Division marked the event, which commemorated the twenty-fifth anniversary of the

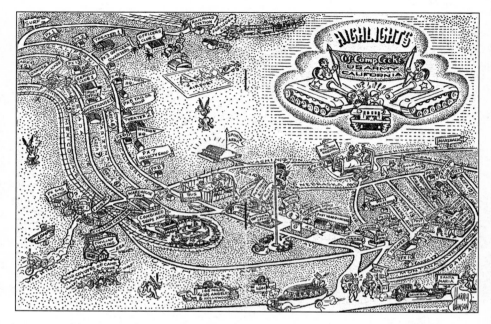

**Major roadways and facilities at Camp Cooke are illustrated on the cover of *Highlights of Camp Cooke U.S. Army California*, a booklet the Army distributed to soldiers at the camp in 1945. Author's collection.**

United States' entry into World War I, with an open house at the camp. The festivities attracted up to 10,000 visitors to the camp for displays, an inspiring demonstration of new Army equipment, a parade of motorized vehicles and marching formations and band concerts. With soldier escorts, guests toured barracks, mess halls, recreation centers, and other buildings. The event opened at 10:30 in the morning and continued through the late afternoon.[10]

Another type of guest house that didn't require advanced reservations and was established exclusively for soldiers who committed serious offenses was the Post Guard House along Kansas Avenue, and a large Post Stockade at 18th Street and Nevada Avenue. Additionally, each Army division at Cooke during World War II, starting with the 5th Armored Division, operated its own stockade.[11]

Shortly after the attack on Pearl Harbor, Japanese submarines began operating off the West Coast of the United States. They struck five merchant vessels, sinking two off shore from California in late December 1941. Rumors began to circulate of an impending Japanese raid on the U.S. mainland. This notion was reinforced on February 23, 1942, when a Japanese submarine rose out of the sea off Goleta, forty miles south of Camp Cooke, and fired about thirteen shells from its deck gun into the Ellwood Oil Refinery. Though the physical damage was negligible, the attack added to the fears of an already jittery population. Many, including the Army, considered the shelling as a diversion for an impending major air strike at a larger target.[12]

On the night of February 24–25, 1942, Army reports of Japanese aircraft over Los Angeles and as far south as Long Beach set off a barrage of antiaircraft guns in what became known as the "Battle of Los Angeles." Within an hour, gunners had fired over 1,400 rounds of ammunition into the air at imaginary "targets." In all the excitement, the only casualties were one death from heart failure, traffic accidents caused by blacked-out streets, and shell fragments from the artillery barrage falling on automobiles and other property.[13]

The attack on Pearl Harbor and a general fear of reprisal after Jimmy Doolittle's bombing raid on Japan on April 18, 1942, hastened the Army Signal Corps' IV Interceptor Command to establish secret early warning radar reporting units along the West Coast between Canada and Mexico to provide air defense. One of the first of sixty-five radar stations built during the war was set up just south of Camp Cooke on Sudden Ranch at Red Roof Canyon in March 1942. Initially designated as Station B-7, it was near the tiny settlement of Arlight, which derived its name from the nearby lighthouse at Point Arguello. About two miles south of the lighthouse was the Coast Guard Rescue Station and Lookout Tower, established in December 1936.[14]

The mobile, trailer-mounted SCR-270 radar system was positioned on a slope overlooking the ocean. The station was manned by a platoon of about fifty men from the 654th Signal Air Warning Company (Frontier), under 2nd Lt. Emil O. Lindner until March 1943 when he was succeeded by another lieutenant.[15]

During the first few months at Red Roof Canyon, the men bivouacked in tents. Later in the year, small wood-framed barracks and an operations building were constructed in a sheltered grove. Much of the radar equipment was installed in an underground concrete vault. The station operated in shifts, 24 hours a day, and communicated by direct phone line to the Los Angeles Air Defense Wing at Eighth and Flower Streets in Los Angeles. The entire station was enclosed behind a wire fence and protected by a detachment of armed military police.[16]

In January 1943 the station was reassigned to the 658th Signal Air Warning Reporting Company and renamed B-30. It is unclear whether B-30 disbanded in June 1944 and the site cleared of all buildings, as reported in some documents, or continued operating until the end of the war.[17]

Soon after the Army activated Camp Cooke, it began hiring large numbers of civilian employees. They filled critical jobs that kept the camp operating efficiently from day to day, and were used in line with War Department policy to relieve military personnel for combat duties. Civilian positions ranged from administrative and clerical to the industrial trades. Included in the latter group was the camp's fire chief. Many of these employees had family members and sweethearts fighting overseas, and all were eager to do their part for the war effort.[18]

In 1944, Camp Cooke had its own version of the iconic "Rosy the Riveter." They were a group of female "oil checkers" who were hired to replace their male enlisted counterparts, formally assigned to the Camp Quartermaster's fuel oil distribution section. This oil was used to heat living quarters and buildings on the post.[19]

Donning coveralls, these women would climb aboard tanker cars brought into the camp by locomotive, and transfer the fuel to tanker trucks. Each truck, which distributed the fuel around the camp, had a checker assigned to accompany the driver. The checker would record the meter readings before and after the oil delivery, and complete an individual ticket for each truck. Coleen Englehart, the chief checker, tallied the daily delivery tickets and submitted weekly reports to the Camp Engineer.[20]

Civilian employees made up a large proportion of the workforce at Camp Cooke. In the image above, field jackets, shirts, trousers, and other apparel of Quartermaster issue are sized and marked by women employees in the Clothing and Equipment Shop. U.S. Army/*Cooke Clarion.*

Sometimes job vacancies were difficult to fill because of an acute housing shortage in Lompoc caused by the opening of Camp Cooke and a lack of transportation, especially for those working night shifts, split shifts, and all types of odd shifts. In late 1942 the Army began to address the problem by building a two-story dormitory to house critically needed telephone operators and other women employees of the post signal office. The Associated Telephone Company furnished a car to be used as a taxi for telephone operators. Vehicle access assured that operators met their work schedules, and transportation was available for additional operators during peak demand periods.[21]

The success of the first civilian housing unit at Cooke led to the construction of additional dormitories for civilian employees eligible for government housing. On October 1, 1943, construction started on three dormitories and one community recreation building in what became known as the civilian war housing area at Utah and Iceland Avenues. The new single-story buildings were of standard Army barrack design and housed up to forty-eight people each. They were built by Harris Construction Company under an Army contract for $89,480.[22]

By February 1945, the civilian housing area included an equal number of twelve separate dormitories for women and men, a recreation building, and a cafeteria. The cafeteria was open Monday through Saturday to serve three meals a day. Whenever possible, the Army also provided transportation to and from the job sites for many of its civilian employees residing at these units.[23]

In the local community, relief from the housing shortage was aided by the federal government's Defense Housing Program. Its initial effort in April 1942 was a trailer camp for up to 200 trailers on twelve acres of land in Lompoc. The government trailers were available to defense employees at Camp Cooke and for families of non-commissioned officers of the first three enlisted grades. Each trailer could accommodate a family of four, and rented for approximately $6 to $7 a week.[24] Later that year the Federal Public Housing Authority approved the construction of two civilian housing developments in Lompoc. The first one was called Lompoc Gardens and consisted of forty units between I and K streets. Built by the Martin & Jepsen Company of Santa Monica, it was completed in October 1942 for about $190,000. The second development, a temporary project of 160 units at Ocean Avenue and O Street, was named Lompoc Terrace. It was completed a year later, in September 1943.[25]

Because Camp Cooke had substantially increased the civilian population of Lompoc, provisions in the defense program required the federal government to meet the community's hospital requirements. In January 1943 the government funded construction of the Lompoc hospital at D Street and Hickory

Avenue. In August 1946, after the camp had closed, the government sold the 40-bed facility on a payment plan to the City of Lompoc.[26]

In 1943 the tide of war began to turn in favor of Allied forces. In May, North Africa was cleared of all German-Italian forces. On July 10 the Allies invaded Sicily and then the Italian mainland at Salerno on September 9. On the eve of its invasion, Italy surrendered. The Germans disarmed its former ally and began taking up defensive positions in the peninsula. The fighting turned into a bloody struggle, with heavy losses on both sides, that continued up to the surrender of Germany in May 1945. On the Russian Front, the Germans were defeated at Stalingrad and at Kursk, marking the beginning of a long German retreat from the Soviet Union. In the Pacific, the Americans were subjecting Japanese-occupied islands to merciless bombings and landed troops on the Solomon, Ellice, and Gilbert Islands.[27]

January 1943 also brought the 2nd Filipino Infantry Regiment to Camp Cooke. It was activated at Fort Ord on November 21, 1942, to accept scores of volunteers that had already filled the ranks of the 1st Filipino Infantry Regiment established earlier in the year at Camp San Luis Obispo. Around the middle of 1943, about three-hundred men from the 2nd regiment were sent to

**M-3 tank destroyers of the 6th Armored Division let loose with a volley. Courtesy the Lompoc Valley Historical Society.**

Australia for covert missions against the Japanese in the occupied Philippines. Earlier in the year a law was enacted that exempted men over the age of 38 from having to serve in the military. Many of the older soldiers in both regiments opted to leave. To help bring the 1st regiment up to strength before it departed to the Western Pacific in April 1944, soldiers were reassigned from the 2nd regiment. On March 27, 1944, the depleted 2nd regiment was disbanded (less one battalion) and was concurrently redesignated the "2nd Filipino Battalion (Separate)" under Lt. Col. Edwin Sallman. He succeeded Col. Charles L. Clifford, who had commanded the unit since its activation. The battalion departed for New Guinea on May 31, 1944, where it was assigned to the Army's Counter Intelligence Corps. It later moved to the Philippines and was inactivated on December 21, 1945.[28]

On March 20, 1943, three days after the 5th Armored Division departed from Camp Cooke, the 6th Armored Division rolled in from the Desert Training Center in California. The "Super Sixth" trained at Cooke until January 1944. Seven months later, on July 19, it became the first combat unit from Camp Cooke to land in France.[29]

The training of ground combat troops at Camp Cooke took many forms and varied in scope that ranged from toughening the individual soldier to maneuvers involving large formations of men and mechanized vehicles during daylight and at night. Much of the training consisted of live firings on ranges or in simulated combat situations. At other times the training was specialized to the unit, such as bridge building for engineers. One unnerving requirement of all soldiers in armored units was to crouch down in a foxhole while a tank rolled over the concealed position. On occasion large exercises included coordinated ground and air operations. The training program at Cooke and at other installations was determined by Army Ground Forces training directives.[30]

One of the earliest courses constructed by Army engineers at Cooke in 1942 was an obstacle course in Oak Canyon. Improved over time, it initially extended for a mile and a quarter and included rope swings, a horizontal ladder, climbing walls, a tunnel, duck walking, numerous simulated bomb craters, wading ponds, and concealed trip wires. Soldiers would complete the course running up the dell to the shale pit carrying a pack with full field equipment. Other parts of the early training program included weapons familiarization, range firings, physical fitness, and long marches.[31]

By early 1943 the infiltration course, close combat firing course, and a combat-in-cities fighting course were introduced into the training program to acclimate trainees to the sights and sounds of a battlefield environment. Collectively they were often referred to as the "battle inoculation series." The main objective of these battle courses was to introduce greater realism into the training, based on the combat experiences that ground forces were beginning to acquire with the landings in North Africa and in the Pacific.[32]

The 6th Armored Division was the first unit at Cooke to complete the infiltration course. The course extended about 100 yards and required troops to crawl across broken ground, and under and around barbed-wire entanglements, with machine gun bullets whistling closely overhead and explosive charges of nitrostarch set off around them. After passing though this zone, the men dug foxholes and fired on the enemy machine gun, destroying it. They then charged dummies with bayonets. In the close combat firing course, trainees walked through a designated avenue of rough terrain and fired at targets that popped up unexpectedly at different distances around them. The men worked in squads, with a sergeant directing their actions through the entire length of operation. In late 1944 it was replaced by a more realistic exercise that featured the "buddy system." Also set in broken terrain with pop-up targets, teams of three or four men would move forward, with each man alternately filling the role of the soldier moving forward and the soldier providing cover.[33]

**Infantrymen of the 6th Armored Division training for urban combat at Mock Village. Sixth Armored Division booklet *A Pictorial History of the Super 6th*.**

**The "town hall" sat at the far end of Mock Village. Sixth Armored Division booklet** *A Pictorial History of the Super 6th.*

The combat-in-cities battle exercise was intended to train soldiers in the proper techniques for capturing a European town that was expected to be rigged with enemy booby traps and snipers. A mock village with some thirty-seven buildings, each structure representing a hidden danger, was constructed in the northwest section of the camp. Training at Mock Village, frequently called "Nazi Village," involved squads of infantrymen conducting house-to-house searches while trying to avoid booby traps, and firing with live ammunition at pop-up targets. The targets were silhouettes that were hinged to doors, windows, staircases, and roofs. They were operated by wires and pulleys. Soldiers who stepped on a booby trap might be sprayed with white flour to simulate an explosive going off. Amid the gunfire, explosives were detonated to simulate artillery shells and land mines. The exercise ended when the Nazi swastika flag flying over the courthouse at the far end of the road would be torn down and an American flag raised in its place. The training village was constructed by an engineering company of the 6th Armored Division and was used by subsequent divisions and later during the Korean War.[34]

Each unit that trained at the village carved into the structures its version of the ubiquitous "Kilroy was here." Scrawled on the side of one building was "McCrary's Snake Emporium." McCrary was a member of the outfit that built the village and reportedly had a reputation of being terrified of snakes. There was the "Harvey Hotel" for an anonymous soldier, the "Bund Club," a reference to the Nazi government, and many more signs like "Hitler doesn't live here anymore" conspicuously displayed.[35]

In late 1943, night fighting was added to the combat training program at Cooke. It was intended to simulate actual conditions that soldiers might encounter on the battlefield. For the armored division, whether conducted in daylight or at night, it involved coordinated operations between tank crews, infantrymen, engineers, and artillerymen. In a typical scenario, the armored forces would destroy an enemy minefield and fire live ammunition while moving forward against simulated enemy positions. After capturing the enemy, the troops would consolidate the area against a potential counterattack. A variation on this type of training had infantry making the assault in advance of the tanks while carefully coordinated artillery and mortar fire passed overhead. During night exercises, whitewash was splashed on the rear of the tanks, and white cloth markers were worn on the backs of each soldier so they could be seen through the inky blackness.[36]

In a four-day exercise conducted under simulated battle conditions in October 1943, the 6th Armored Division's Maintenance Battalion demonstrated their ability to keep up the repair and maintenance work required of them. The battalion moved out to a remote section of the reservation and set

A detachment of 6th Armored Division infantrymen follow closely behind a medium tank as they move up to the "battle front" during night exercises. Notice the whitewash painted on the rear of the tank and the cloth markers on the backs of the soldiers so they can be seen through the inky blackness. U.S. Army/*Cooke Clarion*.

up their equipment at four or five predetermined bivouac areas. Word was then sent to each company in the field that repair work was needed at different locations in the camp. Crews drove to those places, and instead of towing or carting the job to their shops in the motor park, they brought it to their respective bivouac area and did the work under blackout conditions.[37]

In addition to field maintenance, each night, with the exception of the first night, they packed up their equipment and moved from one bivouac area to another under blackout. The exercise was planned so that when changes were made from one bivouac to another the companies would cross trails. The movements were timed to have one company move in just as another company was leaving. At the arrival in each area, equipment and camouflaged vehicles were dispersed, and a vehicular outpost guard was immediately set up.[38]

For ten days, beginning on September 7, 1943, swimming practice and instruction was held at Santa Barbara's Los Baños del Mar municipal pool and at a portion of nearby West Beach for the 6th Armored Division's reconnaissance section. Some 1,300 men underwent rigorous aquatic tests to determine their ability. The tests included swimming 60 yards in six minutes or less while wearing GI shoes and coveralls, and carrying a wooden rifle equivalent to the weight of an M-1.[39]

The aquatics exercise was linked to a military show and the opening of the Third War Bond rally in Santa Barbara. Two detachments of soldiers from the 6th Armored Division with their mechanized vehicles paraded down State Street and encamped at Cabrillo and Dwight Murphy fields. At Pershing Park, Army vehicles were on display and a mock battle scene was staged. Tanks fired blanks from their 37mm guns and machine guns, while exploding whistle bombs added to the sounds of combat. Santa Barbarians were encouraged to buy war bonds and stamps. To help promote bond sales, tank rides were auctioned in De La Guerra Plaza. Before the end of the year, the 6th Armored Division would encamp at Santa Barbara four more times.[40]

Back at Camp Cooke, another form of aquatic training involved stream crossings using different types of floats or rafts. In combat situations, these flotation devices were sometimes used to carry equipment, weapons, and ammunition to an enemy-held riverbank, usually under cover of darkness. These exercises were practiced on the lower part of the Santa Ynez River. Along another part of the river, members of the 25th Armored Engineer Battalion constructed a temporary bridge using enormous sausage-shaped rubber pontoons. They supported metal runways, over which the biggest armored vehicles were able to travel.[41]

The largest maneuver and combat firings in the first two years of Camp Cooke, and the last in which the 6th Armored Division participated before heading to Europe, was held December 1–5, 1943. Two teams of Special Troops in the camp were organized into opposing sides. Supporting the 6th were the 775th Tank Destroyer Battalion, the 382nd Antiaircraft Artillery Battalion, the 461st Ordnance Evacuation Company, and the 190th Ordnance Battalion. Commanded by Maj. Gen. Robert W. Grow, they were probably called the blue team. A red team, consisting of the 2nd Filipino Infantry Regiment, the 606th and 607th Tank Destroyer Battalions, the 156th Engineer Combat Battalion, and the 170th Field Artillery Battalion, led by Col. Harry L. Reeder, opposed them.[42]

The exercise was directed by Headquarters Army Ground Forces and put to the test by Lt. Gen. William H. Simpson, commanding general of the Fourth U.S. Army, with headquarters at the Presidio of Monterey, California. The 6th Armored Division and the various Special Troops stationed at Cooke were all under the Fourth Army, which had recently separated from the Western Defense Command. Chief umpire for the exercise was Col. Wade Gatchell, who headed a 100-man umpire staff from the 13th Armored Division, which at this time was in training at Camp Beale, California.[43]

At the start of the exercise the division's team was massed in the artillery area south of the Lompoc-Surf Road, while the "enemy" red team was entrenched near Tangair on the northern half of the camp.[44]

The first two days of the exercise was devoted to reconnaissance. On the evening of December 2 the blue team moved north across the Santa Ynez River. Forward elements of armored and infantry units launched a limited attack just north of Tangair Road. There the red team had erected silhouette and personnel targets in a fortified area that included mines, barbed wire, and gun emplacements. It then withdrew before the initial attack by the blue team. The 6th Armored Division pulled forward and launched a ferocious assault on the fortified area. By the morning of December 3 the blue team had swept across the San Antonio Creek and was northwest of Marshallia Ranch. At this point it executed a combined frontal assault and wide-sweeping envelopments. The envelopments cut off the red team's escape north to Santa Maria, while units moving from the east pushed the red team toward the sea in the Shuman Canyon area. The umpires ended the exercise on December 4. The following day a critique of the exercise was held at the Sports Arena, with Gen. Simpson as the principle speaker.[45]

Religious worship, like recreation and organized entertainment at Cooke, was an important secondary activity that supported the primary mission of preparing soldiers for combat. Camp Cooke had eleven chapels—designated alphabetically A through E, numerically 1 through 5, and the post chapel. This designation system replaced an earlier policy in 1942 in which each chapel was named for an important Army general officer. Christian and Jewish soldiers shared chapel E at the corner of California Boulevard and 6th Street. On June 10, 1943, a special ceremony was held at Chapel E in which Congregation Adat Israel of Lexington, Kentucky, had placed on loan to the Army a Sefer Torah, the most sacred scroll in Judaism. The service was conducted by Chaplain (Rabbi) Albert M. Lewis of Adat Israel, assisted by the post chaplain at Camp Cooke, Maj. Irving G. Wickman, and Lt. Louis H. Krumbein of the Santa Maria Army Airfield. Brig. Gen. Robert W. Grow, 6th Armored Division commander, attended the ceremony.[46]

In 1943 the 2nd Filipino Infantry Regiment participated in a series of radio and film news stories originating from Camp Cooke that were directed mostly at the Filipino people in the Japanese-occupied Philippines. The messages were of hope and inspiration, and a commitment to the people that liberation was coming.

Perhaps the most rousing propaganda event involving the regiment occurred on June 17, 1943. With rifles slung over their shoulders, the entire regiment passed in review while flashing shiny new bolo knives, the traditional Filipino weapon. They were a gift from the Los Angeles Chamber of Commerce, which raised more than $18,000 in donations to purchase 4,500 bolos. Among the dignitaries on the review stand that afternoon were the regimental commander, Col. Charles L. Clifford, and Maj. Gen. Basilo J. Valdez, Chief of Staff of the Philippine Army and Secretary of National Defense of the Filipino Government in exile. A physician by training, Valdez had a long history of service to his nation. Also on the review stand was Albert Rebel, former business leader in the Philippines and chairman of the chamber's committee that purchased the knives. In brief remarks to the soldiers, Rebel stressed the hope that the Filipino people would one day be free again and expressed his confidence in these fighting men to make that hope a reality. In Tagalog he told the men, "I will meet you in Manila!" They threw up their hands clutching their bolos and cheered. The entire event was recorded on film and in still pictures by all the leading media outlets of the day for distribution across the United States and for shortwave radio broadcasts.[47]

Two months later, three employees from radio station KSFO in San Francisco visited Camp Cooke under the sponsorship of the Office of War Information. The OWI was a U.S. government agency for news and propaganda. The station employees made recordings of about one-hundred officers and men from the 2nd Filipino Infantry Regiment, with each man offering his personal message of patriotism and words of encouragement. Sent across shortwave radio, these messages reached areas occupied by the Japanese. The identities of the soldiers making the broadcasts were concealed for fear of enemy reprisals against relatives and friends of the men. In a second set of recordings, made in April 1944, four representatives from OWI worked with a smaller group of Filipino soldiers and twenty 11th Armored Division soldiers of Chinese heritage.[48]

Customers waiting their turn at the Bank of America. June 1944. Courtesy the American Banking Association (ABA) *Banking Journal*.

Personal banking services came to Camp Cooke on July 26, 1943, with the opening of a Bank of America on California Boulevard in building 11016. The bank was a small sub-branch of its Lompoc office, with V. E. Cordier as director and a staff of two tellers, Christine Gould and Harriet Mortonson.[49]

On September 25, 1943, Col. Carle H. Belt, the second post commander at Camp Cooke, retired from the Army, having reached the statutory retirement age of sixty earlier in the month. Official separation occurred five days later. His successor was Col. Harry C. Brumbaugh.[50]

One of the first decisions made by Col. Brumbaugh was to approve on September 29 an emergency request from Santa Maria Valley farmers that allowed soldiers on special Army passes to harvest tomatoes for up to three days. A shortage of agricultural laborers in the area had put the harvest at risk. Soldiers received the prevailing wage in the district on a per box basis and could earn up to $13 per day. Accommodations were set up south of the city limits for a mobile kitchen and dining room, operated by the State Food Production Council. A large tent served as sleeping quarters for one-hundred men, with an additional fifty spaces available at the USO dormitory established earlier in the year in Santa Maria. The emergency labor shortage ended in early November, with an estimated five-hundred soldiers having worked from one-half to three days in the fields.[51]

The arrival of the first Women's Army Corps (WAC) personnel at Camp Cooke made big news across the camp in 1943. Leading the group of arrivals on September 30 was 2nd Lt. Dorothy Pace, who came from Fort Oglethorpe, Georgia, and was assigned to the Military Personnel Office. Two days later she was joined by 2nd Lt. Clare L. Ventura. On October 15, three more officers

The ceremonial groundbreaking for WAC Officers' quarters at the corner of Utah and Kansas Avenues in December 1943. Shown above is Lt. Dorothy I. Pace driving a nail into the framing for concrete. Watching from left to right are Lt. Beatrice Athanas, quartermaster property office; Col Harry C. Brumbaugh, camp commander; Lt. Clare L. Ventura, civilian personnel office; Lt. Col. William S. Fowler, camp executive officer; Lt. Elizabeth P. Hoisington, PX office; Lt. Eileen M. Knowler, post ordnance shop; Capt. John M. Wyman, director, personnel division; and Lt. Masie B. Davis, sales commissary. U.S. Army/*Cooke Clarion.*

arrived at the camp. They were 1st Lt. Beatrice Athanas, 2nd Lt. Elizabeth P. Hoisington, and 2nd Lt. Masie B. Davis. Several additional WAC officers moved in during the winter of 1943–1944. Privates Virginia Booth and Leota M. Barnett, both from Fort Des Moines, Iowa, arrived at Cooke on April 24, 1944. Probably that same day the WAC detachment was activated.[52]

The WACs set up quarters with station hospital nurses until early 1944, when the officers moved into a new building at the corner of Utah and Kansas Avenues. During this period there were never more than forty-six WACs on the roster, and they worked at the hospital and in various offices around the camp. Lt. Gayle J. Bunts was the commanding officer through most of the detachment's life. On April 7, 1945, Pfc. Grace Funk was the last enlisted WAC to depart Camp Cooke. Closing out the unit and departing three days later was Lt. Grace I. Miller, who had served as an engineer troop supply officer and assistant fire marshal.[53]

Four months later the Army authorized a new WAC detachment at Camp Cooke, which was established shortly after the arrival of Pvt. Sarah Singletary on August 9, 1945. A steady stream of WACs soon followed, coming from general hospitals throughout the Ninth Service Command. Capt. Charles E. Hughes, commanding officer of the hospital medics, was in charge of the WACs until August 19 when Capt. Katherine J. Cargill arrived and took over the job of forming a WAC detachment that eventually numbered some two-hundred personnel before it was inactivated in 1946.[54]

To help finance World War II, the federal government encouraged all Americans to buy Victory Bonds. The campaign appealed to their sense of patriotic

For assisting in a recent bond drive, these seven civilian employees at Camp Cooke received a letter of commendation from the commanding general of the Ninth Service Command. From left to right they are Vera Willett, quartermaster laundry; Edna Simpson, camp headquarters; Lorraine Chilson, quartermaster property; Beth Harding, clothing and equipment; Mildred Perkins, station hospital; Mary Belle Heimann, military personnel; and Leslie Terral, camp engineer. Second Lieutenant Joan E. Andrews, WAC, is shown presenting the letters on behalf of Col. Brumbaugh, camp commander, who also sent letters of commendation to the above women. April 1944. U.S. Army/*Cooke Clarion.*

obligation and democratic principles, feeling that every American should to do their part to help win the war.

In 1942, during the first of several successive nationwide bond drives, civilians and military members at Camp Cooke were urged to participate in the War Department's Payroll Reservation Plan. While civilians were asked to authorize 10 percent of their salaries each payday toward the purchase of a bond, no such requirement was set for military personnel. In all cases, bonds could be purchased separately, but the genius of the Reservation Plan was that it whipped up friendly competition among work sections to determine who could raise the most money. This was successfully demonstrated in the Fourth War Bond Drive that ended in early 1944. Civilians had exceeded the goal of 90 percent participation for at least 10 percent of their pay in the Reservation Plan, entitling Camp Cooke to fly the Treasury Department's "T" flag.[55]

To spur the sale of war bonds in civilian communities, soldiers from Camp Cooke marched in numerous parades and provided marching bands, gave weapons demonstrations, offered rides in military vehicles to those purchasing bonds of large denominations, and did whatever gimmicks were necessary. With assistance from the 69th Armored Regiment, a student bond rally held at Santa Maria Union High School on May 26, 1943, raised several thousand dollars in sales. Soldiers visiting the school presented an informative and entertaining two-part show. An hour-and-a-half morning session in the auditorium included musical selections by the regimental band and singers. Between songs, Army musicians strolled through the aisles playing several accordion and violin solos, especially for the benefit of the women in the audience. Among the guest speakers were two officers and three sergeants recently returned from Guadalcanal who told of conditions on the island, and relayed a few facts about the Japanese soldier. Several captured Japanese weapons were on display.[56]

The afternoon session held in the athletic field included an exhibit of armored vehicles, peeps, a seep amphibious car, and several guns. Watching from the bleachers, the spectators cheered and applauded when soldiers fired two 37mm antitank guns, a 30 caliber machine gun, and other weapons using blank ammunition. An added touch of combat realism was the use of smoke pots to create a smoke screen and the firing of inert rockets.[57]

Both shows were open to students, faculty, and townspeople. Admission to the morning session was a ten-cent war stamp, and a 25-cent stamp entitled one to observe the afternoon demonstration. Purchasers of bonds received rides in the vehicles on display at the athletic field.[58]

Probably the largest gathering and bond event of its type in Santa Maria occurred on August 4, 1943. Military units from Camp Cooke, the Santa Maria Army Air Base, Hancock College of Aeronautics, and the U.S. Navy joined with patriotic organizations in the city for the "All Nations' Festival and Bond Fiesta," with the object of raising enough funds in the county for the U.S. Navy to build a submarine chaser bearing the name of Santa Barbara County. Sponsoring the extravaganza was the Women's Division War Finance Committee. Its members included the Native Daughters of the Golden West, the American Legion Auxiliary, the Navy Mothers of America, and the Order of Pocahontas.[59]

The bond event kicked off at 1:00 p.m. with a parade that started at the high school on Broadway and ended at the Veterans Memorial Building on West Tunnel Street. Marching in the procession were the 69th Armored Regiment band, the 2nd Filipino Infantry Regiment band, and the Santa Maria Army Air Base band, followed by jeeps, half-tracks, tanks, and artillery pieces from Camp Cooke. A 26-foot-long float of the heavy cruiser Los Angeles represented the U.S. Navy. Various other colorful floats from community organizations, and decorated automobiles with community leaders and military officials, were also in the parade.[60]

Speaking briefly to approximate 2,000 people at the Veterans Building were Mayor Marion Rice of Santa Maria and other officials. The festivities were followed by an auction of donated items. The buyers received the articles by purchasing in bonds the amount of the bid. Everything was sold, from a bicycle tube for $25 to a Hereford bull for $45,000.[61]

After a dinner recess the auction continued with an evening of musical performances and vaudeville-style comedy skits by Cooke soldiers, and dancing by all the attendees. Edgar Kennedy, star of stage and screen, acted as master of ceremonies for much of the event and drew plenty of laughs with his "slow burn" act. Toward the end of the evening, Mayor Rice crowned Lorraine Gerlick of Santa Maria queen of the All Nations War Bond Festival. The event ended around midnight and raised approximately $222,000 in the sale of war bonds and stamps.[62]

Overseas, 1944 opened with an Allied amphibious landing at Anzio, south of Rome on January 22. The invasion force attempted to break the stalemate in Italy, but quickly bogged down for the next four months. Rome was finally taken by the Allies on June 4 after the Germans had abandoned the city.[63]

Throughout the year, American and British air forces pounded Germany, although paying a heavy price in crews and aircraft lost. In the Pacific, an American amphibious force came ashore at Kwajalein in the Marshall Islands on January 31, 1944, and on Saipan in the Mariana Islands on June 15. The Japanese effort to resupply its forces on Saipan was also intended to draw the

U.S. Navy into a trap around the Mariana Islands. All aspects of the battle plan were foiled, and the foray turned into a devastating defeat for the Japanese at the Battle of the Philippine Sea on June 19–21. On October 20, U.S. troops secured the beachhead at Leyte at the start of the Philippine campaign.[64]

The biggest war news of the year was Operation Overlord, the Allied invasion plan to liberate Western Europe. On June 6, 156,000 troops, 11,590 aircraft, and more than 6,400 ships and landing craft participated in the invasion along the Normandy coast of France, between Le Havre and Cherbourg. Known as D-Day, the five landing sites were code-named Utah, Omaha, Gold, Juno, and Sword. The American forces landed on the first two beaches while British and Canadian troops attacked from the latter three beaches.[65]

As Allied forces broke through the German beach defenses and swept inland, the main forces pivoted eastward, moving inexorably toward Germany. On August 25, Paris was liberated, and by the end of the year the Allies were closing in on Germany.[66]

Back at Camp Cooke, the 11th Armored Division, with Maj. Gen. Edward H. Brooks, detrained at the camp on February 11, 1944. An advanced detachment arrived a few days earlier, led by Brig. Gen. Charles S. Kilburn. Both groups had come from the Desert Training Center. On March 22, Kilburn succeeded Brooks as the division's commander and a week later assembled the entire division on the parade grounds. In a rousing address to the men he told them they would be ready for combat when their turn came, and announced the selection of the unit's nickname: "The Thunderbolt Division." He also told the men their goal was to "strike the enemy with the power and fury of a thunderbolt." In September 1944 the division left Cooke and landed in Europe the following month.[67]

On March 30, 1944, Gen. Kilburn hosted Postmaster General Frank C. Walker, a member of President Franklin Roosevelt's cabinet, and Maj. Gen. John Millikin, III Corps commanding general, for a medal pin-on ceremony. Walker, who was visiting the camp as a guest of Gen. Kilburn, pinned the Soldier's Medal on the chest of Pfc. Nicholas Dalompinis of the 2nd Filipino Infantry Battalion for heroism in attempting to save the life of Sgt. Santos Bisquerra on January 9, 1944. While out with a group of soldiers at Cooke, Bisquerra was swept from a rock by a high wave into the ocean and knocked unconscious. Dalompinis managed to swim through the rough seas and grab hold of the soldier until he was torn from his grasp by murderous swells.[68]

Army Chief of Staff Gen. George C. Marshall visited Camp Cooke on May 2, 1944, during a 10-day inspection tour of Army installations around the United States. At Cooke he inspected the 11th Armored Division's training

**With duffle bags in hand, arriving 11th Armored Division troops enter their barracks at Camp Cooke. February 1944. Courtesy the 11th Armored Division/Bob Peiffer.**

program, visited the ranges, infantry problem areas, and maneuver sections, and attended artillery firing demonstrations. Fortunate for the division commander, Gen. Marshall was gone when a few days later an artillery practice firing went terribly wrong and injured two employees of the Southern Pacific Company.[69]

The accident occurred on May 12, 1944, during the practice firing of 105mm howitzers on an artillery range by the 492nd Armored Field Artillery Battalion, 11th Armored Division. The impact area of the range bordered the Southern Pacific Company's main railroad line and right-of-way. At about 12:05 p.m. a streamliner "Daylight Limited" passenger train was traveling northbound adjoining the impact area when a high-explosive shell traveled 2,000 yards beyond its target and passed over the top of the train before exploding between the main tracks and the beach. The burst riddled the kitchen car with approximate 32 steel fragments, wounding two women employees and narrowly missing several other crewmembers. The shrapnel injured Alice Johnson in the face and hand, and severed the right arm of Vertie Bea Loggins about six inches below the shoulder. Both women were evacuated at the depot in San Luis Obispo and taken by ambulance to the San Luis Sanitarium. Loggins, the more critically injured, was treated for shock, and later

Camp Cooke as seen looking north. Hundreds of 11th Armored Division soldiers line
the parade grounds between division headquarters (left) and Service Club No. 1 (right)
to hear from their commander. Courtesy the 11th Armored Division/Bob Peiffer.

that day, when her condition stabilized, underwent a reamputation of her arm and the suturing of a six-inch chest laceration. Loggins remained hospitalized until June 15, 1944, and was then transported to her home in Los Angeles.[70]

The 25-year-old Loggins was employed as a dishwasher with Southern Pacific, earning an average monthly salary of $121.90, and was partially supporting her parents. After the accident, Southern Pacific paid her medical and hospital expenses, and provided twenty-two post-operative treatments with the company's surgeon through the end of August 1944. However, the company disclaimed all liability for her injury.[71]

Loggins retained attorney Willedd Andrews, former legal counsel for evangelist Aimee Semple McPherson and founder of the Foursquare Church. Andrews submitted a $50,000 compensation claim with the War Department on September 5, 1944. It was disapproved on January 2, 1945, on the grounds that administrative settlements of claims for personal injury were limited to medical and hospital expenses. Based on advice received from Congressman Clyde Doyle of the Claims Committee in the House of Representatives, Andrews contacted Democratic congresswoman Helen Gahagan Douglas on September 29, 1945. Douglas represented the district in which Loggins legally resided. Douglas quickly took up the claim with the War Department and the Claims Committee. On October 24, she sponsored H.R. 4491, a private relief bill seeking Congressional compensation for Loggins. In preparing the bill, Douglas had reluctantly acceded to the Committee's advice to seek no more than $10,000 in compensation or risk having the bill turned down. When the bill came before the House, Secretary of War Robert P. Patterson submitted a report in which he called the proposed award "somewhat excessive." He recommended a payment of $5,000. Congress accepted the recommendation, and on June 11, 1946, President Harry S. Truman signed the amended relief bill for $5,000 in full settlement with Vertie Bea Loggins. Ten percent of the award went to her attorney.[72]

Loggins, meanwhile, had lost her job with Southern Pacific in January 1945, and because of her disability was unable to find employment for a long time after the accident.[73]

On May 17, 1944, David Abbott "Ab" Jenkins, mayor of Salt Lake City, Utah, and professional race car driver, visited Camp Cooke as technical advisor on automotive preventive maintenance for the Ninth Service Command at Fort Douglas, Utah. During his driving career Jenkins set numerous racing records, some of which remain unbroken, and helped put the Bonneville Salt Lake Flats on the map for auto racing. Jenkins was visiting the command's military installations to help promote the Army's program to conserve gasoline, tires, and parts, all of which were needed for the war effort. At Cooke, Jenkins spoke with groups of maintenance personnel at the camp motor pool, and showed short films about the importance of automotive care for both military and civilian vehicles.[74]

Allied victories in North Africa and Europe resulted in the capture of thousands of German prisoners, eventually numbering 2.5 million. To ease the U.S. Army's burden of having to guard and care for its share of prisoners in the war zone, the War Department had by May 1945 evacuated 371,683 German POWs to more than six-hundred prison camps constructed across the United States. One of these camps was established at Camp Cooke on June 16, 1944, and held about 1,200 prisoners.[75]

The POW camp at Cooke became an administrative headquarters for sixteen branch POW camps located within a 200-mile radius of Cooke. The camps were at Tulare (two), Chino, Goleta, Corcoran (three), Shafter, Lamont, Tipton, Saticoy, Old River, Buttonwillow, Delano, Rankin, and Lemoore. The highest number of prisoners recorded at Cooke and its branch camps was 8,700 on January 1, 1946.[76]

The prisoners were treated in accordance with the Geneva Convention of 1929. To ensure compliance with the accord, representatives from the International Red Cross routinely visited each camp. Prisoners without physical disabilities were assigned to jobs outside the stockades. At Cooke they worked on the military installation at mechanical and civil engineering services, and in clerical positions, the laundry, and even food service. Prisoners also worked on local farms to help relieve the civilian manpower shortage caused by the war. They were paid 80 cents a day, roughly equivalent to the $21 a month paid to an American private in 1941. Prisoners who did not work, or performed administrative and maintenance details within the stockade, received a gratuity of 10 cents a day. This enabled them to purchase items at the POW canteen. Payment in all cases was made in canteen coupons or credited to the individual POW's trust fund account established in the U.S. Treasury Department. A check for the balance of each account was issued to the prisoners upon their repatriation.[77]

Another unexpected guest at Camp Cooke were Italian Service Units, organized into quartermaster service companies. ISUs were created after the surrender of Italy in September 1943 when many Italian prisoners volunteered to work for the U.S. Army. They wore American uniforms bearing a shoulder patch with the word "Italy" written in white letters on a green background. They were paid for their labor, but unlike German POWs, the Italians enjoyed special liberties and privileges that included off-base sightseeing tours sponsored by American citizens. They were kept apart from their former German allies and resided in barracks in the main camp area under nominal guard.[78]

The first Italian Service Unit arrived at Camp Cooke on September 23, 1944. Designated as the 142nd Italian Quartermaster Service Company, it was under the command of Capt. Daniel G. O'Reardon. Two Italian officers, one a captain and the other a lieutenant, accompanied the unit. A second Italian Service Unit, the 140th Italian Quartermaster Service Company, arrived at the camp a month later, on October 24, under Capt. Roderick J. Matthews. Both companies received jobs working in the Combined Maintenance and Camp Utility Shop's machine and automotive sections. They also performed various other labor around the camp as needed. On June 18, 1945, the 142nd departed Cooke for Camp Adair, Oregon. Less than two months later, on August 1, the 3rd Italian Quartermaster Service Company, led by 2nd Lt. William S. Purdy, moved into Camp Cooke from Camp Hahn at Riverside, California. The repatriation of these men began on September 7, 1945, when the 3rd and 140th Italian Quartermaster Service Companies went by rail to Camp Irwin, near Barstow, California, the first stop on their homeward journey.[79]

Attending to the spiritual needs of its Catholic troops, the 11th Armored Division sponsored an outdoor religious retreat at the historic La Purisima Mission, ten miles east of Camp Cooke near Lompoc, on June 23–25, 1944. Nearly 1,000 men camped outside the mission for three days of devotional services. The Reverend John J. Cantwell, Archbishop of the diocese of Los Angeles, officiated at the Sunday morning Pontifical High Mass and an outdoor confirmation service.[80]

In August 1944, several thousand 11th Armored Division troops watched a demonstration of coordinated ground and air combat operations at Cooke by fellow division soldiers and U.S. Navy TBF Avenger torpedo bombers. The targets were two fortified pillboxes that dominated the mock enemy's forward line, backed by an intermediate stronghold. In the opening phase of the attack, the division's 105mm howitzers pounded the main hostile position to the rear while laying down a smoke screen for advancing infantry and tanks. The division's engineers destroyed the pillboxes with bangalore torpedoes, pole charges, and flamethrowers. This created a gateway for tank gunners and the infantry's 57mm assault gun to breach the immediate stronghold. Colored smoke and panels marked the front lines for the supporting aircraft as they joined the ground forces in hammering the main enemy line with bombing and strafing runs. Medics followed the combat forces as they forged ahead under supporting fire, picking up simulated casualties and removing them in ambulance half-tracks to first aid stations. Running commentary was announced over a public address system during the two-hour demonstration. Among the high-ranking officers viewing the exercise were Maj. Gen. John Millikin, III Corps commander; Brig. Gen. Charles S. Kilburn, 11th Armored Division commander;

11th Armored Division tank crews in front of their M-4 medium tanks at Camp Cooke. Courtesy the 11th Armored Division/Bob Peiffer.

With bayonets fixed, two pairs of 21st Armored Battalion infantrymen work their way over double barbed wire at Camp Cooke. Infantry tactics called for first assaulters to throw themselves upon the wire, their weight pulling the barbed strands down so that succeeding waves may get through fast and continue the charge without providing vulnerable group targets. Courtesy the 11th Armored Division/Bob Peiffer.

**Part of the Army's K-9 Corps that demonstrated their canine combat prowess at Camp Cooke during the first week of August 1944. Courtesy the 11th Armored Division/Bob Peiffer.**

Navy Commander P. C. Williams, San Diego Naval Base; Marine Col. J. R. Johnson; and several other Army officers from the Armored Center at Fort Knox, Kentucky.[81]

This coordinated aerial and ground demonstration may have been the first of its kind at Camp Cooke, following the expansion of the existing runway and turnaround ramp to handle larger aircraft. The 676th Engineer Light Equipment Company accomplished the new work in May 1944. The initial runway, built by the 25th Armored Engineer Battalion of the 6th Armored Division and the 155th and 156th Engineer Combat Battalions, was designed for small L-4 "grasshopper" aircraft used to observe artillery fire during training exercises, and to check the effectiveness of camouflaging work from the air. Its construction began in July 1943 with the clearing of brush, and was completed six months later with the paving of two blacktop runways almost at right-angle to each other.[82]

In August 1944 the Army's War Dog Replacement and Training Center at San Carlos, California, sent to Camp Cooke a team of officers and enlisted

**Men of the 56th Armored Engineer Battalion, 11th Armored Division, who received intensive training in handling land mines, are shown distributing a mine pattern at Camp Cooke. Courtesy the 11th Armored Division/Bob Peiffer.**

men with thirteen dogs. The team demonstrated the value of working dogs in detecting minefields, trip wires, and booby traps, and as messengers. Nearly four-hundred officers and soldiers watched as the dogs worked an area holding 180 practice anti-tank and anti-personnel mines, and seven trip wires planted at irregular intervals. The dogs located virtually all the obstacles.[83]

Troops of the 11th Armored Division practiced the concealment and removal of mines, sometimes under menacing sniper fire and the hazard of setting off trip-wire booby traps. All of the men also made at least one trip through Mock Village.

Mustard gas, with all its lethal properties, was added to Camp Cooke's realistic combat training on August 9, 1944, with the opening of the Chemical Warfare Services' familiarization area. It was located in an isolated cove near the San Antonio Creek, several miles north of the main camp gate. The course consisted of a fenced-in field 125 yards square. Inside the square, which was completely devoid of vegetation, were five gas land mines holding fifty pounds of mustard gas. The mines were exploded before troops entered the area during a training exercise.[84]

In August, less than a month before the 11th Armored Division departed for Europe, it celebrated the second anniversary of its activation with a series

Three soldiers of the 11th Armored Division's 21st Armed Infantry Battalion making their first trip through Camp Cooke's mock village fighting course to "clean out" a house occupied (or recently occupied) by simulated enemy troops. The man at the top, who just ascended the stairway with the assistance of the man at the lower left to avoid possible booby traps concealed on the stairs, begins reconnaissance of the building's interior. The rifleman at the lower right cocks his weapon to prepare for first floor reconnaissance. Courtesy the 11th Armored Division/Bob Peiffer.

Batteries A and C of the 491st Field Artillery play a fast game of volleyball. August 1944. Courtesy the 11th Armored Division/Bob Peiffer.

Headquarters and artillery battery men of the 490th Armored Field Artillery fight it out in a tug-of-war. August 1944. Courtesy the 11th Armored Division/Bob Peiffer.

of athletic events. These included softball, basketball and volleyball, and a competition to find who could throw a practice grenade the furthest.[85]

On September 6, 1944, the 86th "Black Hawk" Infantry Division, under Maj. Gen. Harris M. Melasky, began arriving at Camp Cooke from Camp Livingston, Louisiana. Less than a month later, on October 1, the division moved north to Camp San Luis Obispo for seven weeks of amphibious training. It returned to Cooke on November 23. Eleven days later it was on the move again and arrived in France on March 1, 1945. The division used Camp Cooke largely as an assembly point and outfitting center prior to going overseas.[86]

Camp Cooke welcomed the 97th "Trident" Infantry Division, with Brig. Gen. Milton B. Halsey, on September 30, 1944. They arrived about the same time that the 11th Armored Division and the 86th Infantry Division were passing through the camp. Over the next four months the division received intense combat training before landing in France on March 1, 1945.[87]

One example of their training in October 1944 involved an attack on a fortified enemy position. Starting from an area thick with trees and high grass, troops of the 2nd battalion, 386th Infantry Regiment, stealthily moved within range of the enemy pillbox. Armed with rifles, bazookas, bangalore torpedoes, grenades, and flamethrowers, they worked their way through a minefield and blew their way through a barbed wire barrier. During this time, explosive charges placed in the ground were detonated to simulate overhead artillery fire raining down on the troops. When they reached assault distance on the

Advancing on an "enemy" position at Camp Cooke under cover of smoke during a field problem are troops of the 2nd Battalion, 386th Infantry Regiment, 97th Infantry Division. The troops are armed with grenades, bazookas, Browning automatic rifles, and other infantry arms, and are preparing to rush a pillbox after the softening up process had been completed. October 1944. U.S. Army/*Cooke Clarion.*

A composite photograph showing troops of the 97th Infantry Division training. The picture at the left shows Cpl. Ben Phillips, antitank company, coming out of his foxhole to toss an explosive charge at an onrushing tank. At right, Lt. Franklin M. Koons explains to Pfc. Jenous M. Hampton the value of a well-placed rocket from the bazooka in damaging a fortified enemy position. October 1944. U.S. Army/*Cooke Clarion.*

target, smoke pots were released to cover the movement of the lead troops, who rushed forward and tossed explosive charges inside the pillbox. The main force then moved up to clean out the pillbox and to engage dummy "troops" behind the fortification.[88]

In November 1944, religious oblations were held for Catholic, Jewish, and Protestant officers and soldiers of the 97th Infantry Division. The first of these "Spiritual Preparedness" services, organized by the division and camp chaplains, was held on the morning of November 22. More than 5,500 Protestant soldiers jammed the Sports Arena to hear an inspiring message delivered by Dr. James W. Fifield, Jr., pastor of the First Congregational Church of Los Angeles. "God, My Co-Pilot" was the theme of Dr. Fifield's sermon, during which he brought home to the troops the urgent need to get closer to God. Sermon cards were distributed, and the entire assemblage joined in the singing of several well-known hymns. Jewish soldiers filled Chapel E to its 300-seat capacity to hear a sermon by camp Chaplain Rabbi Perry E. Nussbaum, and to recite religious prayers. Both Chaplain Nussbaum and Dr. Fifield told their congregants to be strong and to keep faith with God in their fight against tyranny.[89]

Three days later, on November 25–26, 1944, some 2,500 Catholic soldiers of the 97th Infantry Division and other units at the camp joined in a religious retreat of inspirational sermons and religious services at La Purisima Mission. The National Catholic Community Service of the USO sponsored the event, which was carried out jointly under the guidance of Monsignor Edward R. Kirk, pastor of St. Basil's parish of Los Angeles, and the Catholic chaplains at Camp Cooke. Approximately 147 servicemen received the Sacrament of Confirmation from the Archbishop of Los Angeles, John J. Cantwell. Franciscan Monks from the mission in Santa Barbara furnished the choir and music of the morning High Mass. On the afternoon of the second day, the men marched to the Veterans Memorial Building in Lompoc for a joint civilian-military ceremony that ended the program.[90]

The year 1945 brought Allied victory and an end to World War II. In Europe, January opened to a continuation of a struggle that started on December 16, 1944, when a German counteroffensive through the Belgian region of the Ardennes Forest caught American troops by surprise and created a bulge in the U.S. lines. Outnumbered and in some instances trapped, the Americans

Some 2,500 Catholic soldiers of the 97th Infantry Division and other units from Camp Cooke attended a special religious retreat at La Purisima Mission. November 25–26, 1944. Courtesy the Lompoc Valley Historical Society.

fought tenaciously, while reinforcements from the First Army in the north and the Third Army in the south turned to squeeze off the counteroffensive in what became known as the Battle of the Bulge. Near Bastogne, S/Sgt. Archer T. Gammon of Camp Cooke's 6th Armored Division earned the Medal of Honor for his daring attack on the German line in which he wiped out two enemy machine gun crews, cut down two hostile infantrymen, and advanced on a Tiger tank before being stopped. By January 5, the Germans began withdrawing, and before the end of the month were pushed back to their starting point.[91]

On March 9, American troops crossed the Rhine River at Remagen and, with the British to the north, were reaching deep into Germany. The Anglo-American armies stopped at the Elbe River, fifty miles west of Berlin, leaving the Russians to seize the city on April 30. On that same day, Adolf Hitler and several of his top henchmen committed suicide. A week later, on May 7, German representatives signed the unconditional surrender of all German forces to the Allies at Reims, France.[92]

American forces were also closing in on Japan. In two of the most savage battles of the Pacific war, the U.S. Marines liberated Iwo Jima on March 16, 1945, and Okinawa on June 22. B-29 bombing raids were devastating Japan in the hope that it could be brought to surrender.[93]

As the war moved closer toward total victory for the Allies, the sudden death of President Franklin D. Roosevelt on April 12, 1945, shocked Americans. Memorial services were held throughout the country. At Camp Cooke, soldiers and civilians gathered at Theater No. 3 on Sunday morning, April 15, to honor the late president. Camp chaplain (Chester) C. U. Strait conducted the services that opened with an organ prelude and was followed by the Westmont College Choir of Los Angeles, under the direction of Helen Catherwood Strandberg. The choir of eleven male and twenty female voices sang "Battle Hymn of the Republic," "God Is Marching On," and other selections. The services included scripture readings, violin solos, and a benediction, and ended with the playing of taps.[94]

The choir was on a tour of the Central Coast and performed a short concert at the Station Hospital on Saturday evening. They returned to Cooke Sunday morning for the memorial service then headed to a previously scheduled morning engagement at a church in Lompoc.[95]

Plans, meanwhile, were moving ahead for an invasion of Japan that would involve air, naval, and ground forces in two amphibious assaults. The first landing was scheduled for November 1945, and the second for March 1946. To supplement the invasion force, the 13th and 20th Armored Divisions and other smaller units were withdrawn from Europe and transported to Camp

Reenacted several times was the image shown above in the top picture of troop trains bringing 13th Armored Division soldiers to Camp Cooke. The arriving troops were met by advance detachment representatives of their respective divisional units. Photograph lower left shows the 64th Army Ground Forces band, under the baton of Chief Warrant Officer Marion C. Walter, which welcomed each train. Lower right shows the incoming troops wearing sun tans, which they will exchange for olive drabs during their stay at Cooke, being taken by truck from the detraining area to their barracks. September 1945. U.S. Army/*Cooke Clarion.*

Cooke between July and September 1945 to train as part of the amphibious force. Also sent to Cooke from Fort Ord in July for possible redeployment against Japan was the 764th Amphibian Tractor Battalion.[96]

In preparation for the arrival and training of redeployment forces at Cooke, tons of materials and vehicles began rolling into the camp and filling warehouses and storage yards to capacity. Vehicles of every description, including M-4 tanks, half-tracks, 6x6 trucks, reconnaissance cars, jeeps, and heavy maintenance equipment, were unloaded from railroad flatcars. By the middle of August more than 150 railroad cars of equipment, including 100,000 pounds of ordnance and 1,714 vehicles of all types, were unloaded at Cooke.[97]

In July 1945 post engineers and training personnel with a labor detail from Italian Service Units began constructing three specialized redeployment training areas. They would consist of two mock Japanese platoon defense areas and an amphibious landing area. The landing area was north of Surf, between the coastline and the salt water barrier bridge. The training aids at this site would consist of ship mockup landing crafts to practice surf-to-shore landing operations, a landing tower with nets built on land, and a second netted tower on pontoons in the lagoon below the bridge. Soldiers would use the netted towers to practice disembarking from a ship. The mockup landing crafts would be made from eucalyptus logs with concrete decks, and range in size from 150 to 300 feet long. Site access facilities would include a new shale road, a footpath under the lagoon railroad trestle, and a vehicle parking lot.[98]

The first of two Japanese defense areas was under construction a few miles southeast of the landing area at Lynden Canyon. It was located in what is now south Vandenberg AFB, about 1,000 yards inside the main entrance. The second Japanese defense area was on the northern half of the camp, between Tangair and the existing European-style mock training village. Both defense areas would consist of barbed wire obstacles, concrete and log pillboxes, and simulated Japanese cave fortifications. One of these areas was about 200 yards wide and 500 yards long.[99]

The dropping of an atomic bomb on Hiroshima on August 6, and another three days later on Nagasaki, brought the unconditional surrender of Japan on August 15, 1945, that ended World War II. The Japanese redeployment training areas at Camp Cooke were no longer required, and sometime later they were removed.

With the end of World War II the War Department began a rapid demobilization of its military forces to a peacetime footing. Included in the process was the separation of personnel, the sale of surplus property, and the closing of facilities. American service personnel were eager to go home. Separation for enlisted men started in September 1945, based on a critical score—a point system—requiring 85 points and two years of active service. Also first for discharge were men over 38 years of age. For enlisted women the required score was between 29 and 33, with one year of active service. Within weeks of the initial rules governing separation, the critical score was successively lowered, making for a more rapid separation.[100]

At Camp Cooke the Army established a Separation Point at the Station Hospital on September 15 to speed up the discharge of eligible service members at the camp waiting to be sent to Separation Centers scattered throughout the country. After a hectic three months, a change in War Department policy affecting all Army Service Forces Separation Points closed the local Point, which up to this time had released 8,913 personnel, including thirty-one Army nurses, into civilian life. The closing stranded several hundred men eligible for immediate discharge who would now have to be scheduled and routed to Separation Centers.[101]

The first of a long line of troop trains going to Separation Centers departed Cooke on January 25, 1946, when 260 men pulled out for Fort McArthur in Los Angeles, followed a few days later by additional trains headed to other parts of the country.[102]

During this same transitional period, the 13th Armored Division was inactivated on November 15, 1945. Its remaining personnel, many of whom were waiting for separation orders, were reassigned to the 20th Armored Division at Cooke. Large stocks of equipment and vehicles also transferred to the 20th.[103]

On February 7, 1946, the War Department announced that Camp Cooke would be inactivated the following month. More than anything else, the order erased the uncertainty about an immediate post-war mission for the camp. On March 4–8, troops of the 20th Armored Division entrained for Camp Hood, Texas, where one month later the division was inactivated.[104]

With the departure of Army ground and service forces from Camp Cooke, many of the service facilities began shutting down. Guest House No. 1 closed its doors permanently on January 18, 1946. Closing in March were the second Guest House, Service Club No. 1, two chapels, several post exchanges, the Bank of America branch office, the Red Cross field office, which had operated at Cooke since February 1942, and Theaters 1 and 2. Theater No. 4 closed a few months earlier, leaving open only Theater No. 3. On April 5 the *Cooke Clarion* ceased publication. In the waning days, one function after another closed, including the large Station Hospital. The last of the German POWs moved out of the prison camp a few days before it closed on May 18.[105]

On June 1, 1946, Camp Cooke was officially placed on inactive status. Except for a small caretaker detachment assigned to look after civil engineering and security at the post, the camp was now a ghost town.[106]

# 3

## THE INTERWAR YEARS, 1945–1950

On July 5, 1945, one day after America celebrated Independence Day, the Robert E. McKee Company of Los Angeles, one of the largest general contractors in the Southwest, broke ground at Camp Cooke for the construction of a maximum security U.S. Disciplinary Barracks (USDB) to confine American military prisoners. The prison would consist of eleven three-story concrete housing blocks connected off a main corridor, with an administrative building similarly connected at the center. Included along the hallway would be a control center, the prison infirmary, a mess hall, a movie theater, a chapel, and academic classrooms, all part of the principal structure. Behind these buildings would be a large industrial facility for vocational programs, a laundry, a central heating plant, and a firehouse building, all separate structures, and a recreation field for inmates. The central heating plant and the firehouse were located outside the chain link fence and guard towers that surrounded the main facility.[1]

McKee would also construct several Theater of Operations barracks to house an initial staff of approximately 450 Army guards and fifty officers. These structures were probably removed from the main camp and refurbished. At $4.5 million, it was the largest of three construction contracts awarded for the prison project on June 30. The other two contracts went to Hoagland-Findley Engineering, and Wonderly Construction, both of Long Beach, California. They would build the sewer system, disposal plant, a water reservoir, and the pipeline between the water treatment plant at Camp Cooke and the prison's reservoir. The final cost of the project was more than $6 million. The total size of the disciplinary barracks reservation was approximately 2,950 acres.[2]

The USDB was built to house 1,551 inmates, but with double bunking the population occasionally increased to nearly 3,000. Numbers fluctuated

Aerial view of the U.S. Disciplinary Barracks. Upper left: enclosed within the fence line are the administration building and unit cell blocks and above the recreation field and industries area. Center: staff housing and mess hall area. Lower center: the farm area. The main roadway to the right of the prison grounds is Santa Lucia Canyon Road, which turns into Floradale Avenue as it nears the town of Lompoc to the south. Note the washed out bridge over the Santa Ynez River. About one mile up from the prison the road branched into Camp Cooke (not shown in the picture) and was known as Pine Canyon Road. January 28, 1952. National Archives.

from month to month, but it was generally between 800 and 1,500. Inmates were principally from the Army and the Air Force, and ranged from privates to a former lieutenant colonel.[3]

Staffing for the USDB began in November 1946 with the arrival of Maj. James M. McCarthy, followed a week later by other officers and a guard detachment. On December 6, McCarthy was succeeded by the first commandant of USDB, Col. James W. Fraser, who transferred from the Adjutant General's Office in Washington, D.C. On December 16 the War Department activated the facility as the Branch United States Disciplinary Barracks. From this time forward, the USDB also assumed caretaker and security responsibilities for the entire Camp Cooke reservation while the camp remained inactive. The caretaker detachment established at Cooke when the camp closed on June 1,

1946, was reassigned to the commandant of USDB, who also served as the commander of the camp during periods of inactivation.[4]

As the War Department returned to peacetime requirements in the immediate aftermath of World War II, it underwent significant restructuring. Starting in June 1946 it abolished service commands, and reassigned Camp Cooke from the Ninth Service Command at Fort Douglas, Utah, to Headquarters, Sixth Army Command, at the Presidio in San Francisco. The USDB, which was activated as a branch of the main Army prison at Fort Leavenworth, Kansas, remained a branch but was aligned under Headquarters Sixth Army. On January 15, 1947, it received the designation Area Service Unit 6103. Later that day the first trainload of 242 prisoners transferred into the USDB from Camp McQuade military prison near Watsonville, California.[5]

By September 1947 the USDB was holding 1,500 prisoners, with a staff of 54 officers, 736 enlisted men, and 122 civilians. Many of the civilians were supervisory maintenance workers, or supervising inmates at the vocational farm about one mile southeast of the prison near the Santa Ynez River.[6]

After Camp Cooke closed, the USDB acquired a large cache of surplus

**Conducting a unit count of prisoners at the U.S. Disciplinary Barracks. National Archives.**

items from the camp, including tools, equipment, building materials, barracks, and probably a few vehicles. Whatever salvageable items remained, particularly building materials and entire buildings, went to auction through the Army Corps of Engineers or through its agent, Bosley Wrecking Company. In 1947 Bosley offered to the public entire 20-foot by 100-foot Theater of Operations barracks at Cooke for $475 each on a cash and carry basis. Between 1948 and 1949 the USDB moved three additional barracks from the main camp to the prison grounds and formed a 6,000-square-foot Officers' Club.[7]

A more doleful reminder of Cooke's recent past were the remains of ten German POWs and four Italian Service Unit members that were exhumed on November 26, 1947, and reburied at the national cemetery in San Bruno, California. Most of these men had died at other camps in California and were temporarily interned at Cooke.[8]

With more than 90,000 acres of profitable land sitting idle at Camp Cooke, and local ranchers eager to put it to good use, the Army leased nearly the entire reservation for agricultural purposes. Leases issued in January 1949, and twice more before December, varied for up to five years in duration. Because of concern that tractors or other farm equipment could set off live ammunition laying on the ground, most of the areas were restricted to grazing. Sheep grazing was especially effective, and proved an economic boon to the Army in controlling the growth of grass and other vegetation in the cantonment and building areas.[9]

After World War II, units from the Army's Organized Reserve Corps and the National Guard periodically held training exercises at Cooke. In early July 1950, California Guard troops from the 114th Antiaircraft Artillery Brigade, the 251st Antiaircraft Artillery Group, the 234th Antiaircraft Artillery Group, the 425th Signal Radar Maintenance unit, and the 93rd Army band conducted two weeks of training at Cooke. During this time, and in separate visits, Secretary of Defense Louis Johnson and Lt. Gen. Albert Wedemeyer, commanding general of the Sixth Army, dropped in at Camp Cooke to observe troop training. After these units departed, personnel from the 421st Antiaircraft Artillery Gun Battalion of the Nevada National Guard spent two weeks at Cooke. On July 22, Governor Vail M. Pittman and other officials from the Silver State visited them.[10]

Less than a month later the United States would be embroiled in the Korean War, and Camp Cooke would play an important role in this tumultuous event.

# 4

# THE CAMP IN THE KOREAN WAR, 1950–1953

On June 25, 1950, the North Korean army launched a surprise invasion of South Korea across the 38th Parallel. With superior forces and arms, they quickly overran the Republic of Korea's (RoK) defenses and reached the capital city of Seoul in four days. President Truman ordered Gen. Douglas MacArthur to commit all forces at his disposal to assist RoK troops. In an effort to delay the North Korean advance, MacArthur sent a 540-man task force from Japan that was quickly rolled up by superior numbers of well-armed enemy soldiers. As fresh American troops poured into Korea, they took up successive defensive positions further south.[1]

Meanwhile, President Truman had turned to the United Nations, which condemned North Korea's actions and demanded that its troops withdraw to the 38th Parallel. When the North Koreans continued to advance, the U.N. Security Council passed a resolution urging U.N. members to provide military assistance to South Korea. At the request of the U.N., the United States formed a U.N. command of American and allied forces. Gen. MacArthur was appointed its commander.[2]

By August, a U.S. and RoK defense line was set up around the port of Pusan at what became known as the Pusan Perimeter at the southeast corner of the country. The situation was critical. On September 15, MacArthur launched an amphibious landing at Inchon behind enemy lines that completely caught the North Korean forces by surprise. American-led U.N. forces went on the offensive and by the end of October had advanced to within sixty-five miles of the Yalu River, the border between North Korea and the Chinese province of Manchuria. At this point they encountered groups of Chinese troops. As U.N. forces worked their way further north, mass waves of Chinese soldiers suddenly attacked on November 25, forcing a general withdrawal.[3]

73

The Chinese advance stalled during the first week of January 1951, about forty-five miles south of Seoul. On January 25, U.S. troops counterattacked, liberating Seoul and driving north of the 38th Parallel. By the end of May 1951, the war had reached a stalemate roughly along the pre-war border of North and South Korea. The two sides fought limited engagements, but neither side could force a surrender. Peace negotiations opened in July 1951, and two years would pass before a truce was signed on July 27, 1953.[4]

Back in the United States, on August 1, 1950, President Truman ordered into federal service the 40th Infantry Division of the California National Guard. It was directed to Camp Cooke, with an activation date of September 1. To get the camp ready for its new arrivals, and to establish a "housekeeping" force to operate the installation, Headquarters Sixth Army issued General Orders 100 that reactivated Cooke on August 7 and established Area Service Unit 6014 as the post headquarters organization. Included in the unit's service functions were the headquarters staff agencies, the hospital, the repair and maintenance shops of the post engineer and the quartermaster, and the ordnance and signal sections. ASU 6014 would eventually consist of about five hundred members. The next major step toward fully reactivating the camp occurred on August 16 when Col. Ben R. Jacobs, commandant of the Branch United States Disciplinary Barracks and commander of Cooke, relinquished command of Cooke to Col. Frank R. Williams.[5]

While troops of ASU 6014 were busy clearing buildings, pulling weeds, and unloading shipments of supplies in preparation for the arrival of the 40th Infantry Division, the 13th Armored Reserve Division from Los Angeles was wrapping up two weeks of combat maneuvers at the camp. Previously an Army Ground Forces unit with combat experience in Europe, the division transferred to Cooke in August 1945 and in November was disbanded. A few years later the Army reactivated it as an Organized Reserve Corps unit with personnel from California and Arizona. Its recent training at Cooke had been partially curtailed because much of the camp was still commercially leased for agriculture. Given Cooke's new wartime mission, the Army cancelled the leases on August 31, and by mid–September all livestock were gone from the camp.[6]

During the four years that Camp Cooke remained quiescent, much of the installation had fallen into disrepair. Roads damaged by heavy armored equipment in training periods during World War II needed resurfacing. Telephone and electric lines had snapped in places, disrupting communications. Vacant buildings, some with leaking roofs, broken windows, and littered with debris, became refuges for wildlife. These were just some of the problems plaguing the camp in the early days of its reactivation.[7]

In September 1950 the Army's District Engineer Office in Los Angeles indicated that seventeen contracts awarded for the refurbishment of Camp Cooke would be completed by mid–January 1951. These contracts amounted to approximately $2.9 million. The Army also hired civilian government employees to assist in the renovation and operation of the camp.[8]

The first troops of the 40th Infantry Division arrived at Cooke on September 6. Commanded by Maj. Gen. Daniel H. Hudelson, the division had about 10,000 men in its ranks, drawn from units across southern California, including Battery C of the 981st Field Artillery Battalion from Santa Maria and Lompoc. The battalion was headquartered in Santa Barbara under Lt. Col. Charles Ott, Jr., with Capt. Robert C. Lilley, Jr., of Lompoc and Lt. Walter Hoffman in charge of Battery C. During the next few weeks augmentees, many coming from other Army installations, filled out the division to full combat strength of 18,000 men. Among those volunteering for service with the 40th were pianist and composer André Previn, and actor Tom Brown. Brown had served in the Army during World War II as a paratrooper, and won the Bronze Star and the French Croix de Guerre twice. Three other soldiers of the 40th Infantry Division, Sergeants David B. Bleak and Gilbert G. Collier, and Corporal Clifton T. Speicher, would each receive the Medal of Honor for exceptional valor against a determined enemy in the Korean War. Also volunteering were eighty-four men from Guam who arrived at Cooke on November 28, 1950.[9]

A small but welcome addition to Camp Cooke was the first Women's Army Corps unit at the camp since World War II. The initial cadre of eight enlisted WACs from Fort Sam Houston, Texas, arrived at Cooke on September 17, 1950. Postal Officer 1st Lt. Ann M. McDonnell was placed in charge of the enlisted women, who were organized into WAC Detachment 6014 Area Service Unit. In November, McDonnell was succeeded by Capt. Tessa Jean Blasingame.[10]

In October the 375th Military Police Company, under 1st Lt. John H. Flannery, arrived at Camp Cooke. It replaced the law enforcement portion of Area Service Unit 6014 as the post military police, responsible for internal security at the camp. It also managed traffic control, and the enforcement of military rules and regulations on the post and in the towns where soldiers from Camp Cooke spent their free time. The unit came under the supervision of the post provost marshal and was responsible to the post commander. Before coming to Cooke, the 375th was part of the Organized Reserve Corps in Azusa, California. Many of its men were veteran soldiers, and a large number were police officers from the Los Angeles area.[11]

Beginning in September 1950, the number of troops arriving at Cooke was increasing rapidly. To meet the demand for all types of training activities and personal services, dozens of facilities reopened with remarkable celerity. These included everything from firing ranges to technical shops, the Sports Arena, the two guest houses, the two Officers' Clubs, the two service clubs, the NCO Club, and three theaters. For 24 cents admission, soldiers and their civilian guests could see Hollywood's latest movies at any of the base's theaters. Soldiers and civilians could read about the latest events happening at the camp by picking up a copy of the *Cooke Clarion*, which resumed publication on September 29 after a hiatus of fifty-three months.[12]

When the Station Hospital closed in May 1946, military medical services at Cooke lapsed for seven months until the opening of the Branch U.S. Disciplinary Barracks. During this time, staff members and their dependents relied on local practitioners, or traveled great distances to military facilities. Some relief came with the opening of a small clinic and hospital at the prison, which also cared for the inmate population. In April 1948, after funding became available, the Army established a separate minimum care annex at the former Station Hospital for military dependents and other civilians authorized to receive Army medical treatment. The annex included two wards, one each for male and female patients, and small sections for dental, x-ray, laboratory, outpatient, and other essential medical care. In its first month of operation the clinic delivered its first infant.[13]

The rising camp population and a corresponding increase in the number of medical cases led to the reopening of the Station Hospital as the U.S. Army Hospital at Camp Cooke, and its separation from the disciplinary barracks, on October 6, 1950. The general order that directed this change also expanded the hospital from a 77-bed facility in mid–September to an authorized bed capacity of 450 patients. Four months later, on March 3, 1951, the hospital expanded to a 1,000-bed facility and was renamed the U.S. Army Hospital, Specialized Medical Treatment Center. Much of its specialty was orthopedic and general surgical treatment of Korean War casualties that began arriving at the hospital later that month, and non-combat medical cases from the Far East and European Commands. By the end of the year, the Cooke medical center had become the Army hospital for all of Southern California, and expanded to a 1,350 patient bed facility. Its medical staff treated a vast array of illnesses and injuries, delivered babies, and, when necessary, consulted with leading specialists from the civilian community. The hospital was also responsible for all matters relating to camp sanitation. Given the hospital's large and diverse patient load, the federal government designated it as a Cortisone and ACTH (Adrenocorticotropic hormone) research center. Patients administered these new drugs at Cooke were closely monitored and the treatment results

The U.S. Army Hospital at Camp Cooke, virtually unchanged from its World War II appearance. January 28, 1952. National Archives.

reported to the U.S. Surgeon General. The hospital also included separate medical and physical evaluation boards for discharging military members with service and non-service related disabilities. The only other facility in California at this time to have these evaluation services was Letterman Army Hospital at the Presidio in San Francisco.[14]

Between October and December 1950, five deaths were recorded at the hospital. Except for an infant who died at birth, the remaining four deaths were all enlisted men. Three of the men were killed in automobile accidents and the fourth from a gunshot wound to the chest that occurred at Cooke. In 1951, sixteen civilians and six military members died at the hospital. Nine were newborns and most had congenital abnormalities. Illnesses and one automobile accident took the lives of the remaining civilians. The six military deaths were all trauma cases, one a Korean War gunshot victim and the second an automobile accident. Circumstances for the remaining deaths are unavailable. Hospital reports for 1952 indicate fifty-six deaths, but less than half are classified. Seventeen were traffic accidents, seven were neonatal, and one was a self-inflicted gunshot on the rifle range that claimed the life of 20-year-old Pvt. Elza E. Mitchell.[15]

**Patients and staff at the Camp Cooke hospital auditorium enjoying organized holiday entertainment. Author's collection.**

The hospital received tremendous assistance from the American Red Cross chapters of Santa Barbara County. Its volunteers worked with military social workers for the benefit of patients, and directly with patients and their families. The Red Cross staff at Camp Cooke operated from a two-story recreation building in the middle of the hospital grounds. The building was accessible to patients in all the wards by a series of ramps. With a large auditorium for shows and movies, a music room, a lounge, a kitchen, and various other rooms, the building had the appearance of an intimate hotel. Between the camp Special Services Office, the Red Cross, and individual donations, the recreation building was furnished with radios, magazines, books, phonographs and records, a juke box, a piano, card tables, and an indoor shuffle board game. Working closely with the Red Cross were a staff of dedicated, trained volunteers from the Gray Ladies Society who administered much of the recreation and personal services program for hospital patients. The Gray Ladies recruited from the neighboring communities a steady stream of talented volunteers who entertained hospital patients.[16]

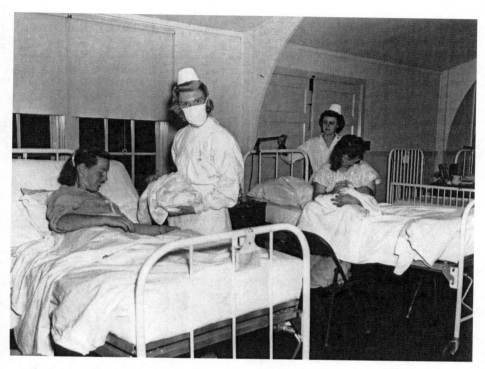

**Nurses assisting wives in the maternity ward at the Camp Cooke hospital. From left to right: patient Mary E. Key, Capt. Eleanor R. McGowen, Geraldine Robtaille, and Lt. Norma J. Linnacke. January 9, 1951. National Archives.**

Other community organizations also volunteered their services to the hospital through the Red Cross. The AMVETS Auxiliary units of Santa Barbara and Santa Ynez provided coffee hour in the patients' lounge twice a month. The American Legion posts in Santa Maria, Guadalupe, and Lompoc took turns hosting weekly Bingo parties. The Native Daughters of the Golden West, the Alpha Club, and the Eagles Auxiliary all gave parties at regular intervals. College groups from Santa Maria and Santa Barbara brought entertainment to the auditorium once a month. Starting in October 1951, the Santa Maria chapter of the Red Cross hosted weekly parties and dances for ambulatory patients at the hospital. Other activities included craft groups, and the showing of movies in wards and in the auditorium. In 1951 the Music Academy of the West near Santa Barbara held a concert performance in the recreation building, led by the baritone of American opera and popular music, John Charles Thomas, and his Romany Chorus.[17]

Republican senator William F. Knowland of California was probably the first member of Congress to visit the camp after it reopened. On October 23, 1950, he met with Col. Frank R. Williams, post commander, and George Miller, president of the Lompoc Chamber of Commerce, to address the local housing shortage.[18]

On the next day, Governor Earl Warren arrived at the camp. The governor was on a tour of military reservations in California, and was welcomed to the camp with a 19-gun salute by the 40th Infantry Division. He later observed a mock tank battle, enjoyed a hands-on antiaircraft demonstration by the 140th Antiaircraft Battalion, and spoke with groups of soldiers.[19]

Later that month more than sixty businessmen from Santa Maria and Lompoc watched a demonstration of weapons firings at Camp Cooke performed by the 223rd Infantry Regiment, 40th Infantry Division. The weapons ranged from machine guns to 155mm howitzers. Five weeks later, on December 1, 1950, the 40th Infantry Division repeated the demonstration for the general public.[20]

**California Governor Earl Warren with Maj. Gen. Daniel H. Hudelson (looking over his left), 40th Infantry Division commander, at division headquarters. October 24, 1950. National Archives.**

In late November 1950, Camp Cooke welcomed the 747th Amphibious Tank and Tractor Battalion, a Florida reserve unit commanded by Lt. Col. Lamar G. Carter. The battalion was activated in 1942 as the 747th Tank Battalion and had participated in the D-Day invasion at Normandy and subsequent fighting in Europe. In March 1946 it was inactivated and later allotted to the Organized Reserve Corps at Gainsville, Florida. Four years later, in September 1950, it was called into federal service as the 747th Amphibious Tank Battalion and ordered to Fort Worden, Washington, for training. Shortly before transferring to Camp Cooke, the battalion added the designation Tractor to its name, becoming the 747th Amphibious Tank and Tractor Battalion.[21]

The battalion was equipped with LVTs, short for landing vehicle tracked. LVT 5s were amphibious tanks, and LVT 4s were amphibious tractors. The former was equipped with a turret fitted with a 75mm howitzer. It also carried a 30 caliber machine gun on the right front of the vehicle next to the driver, and occasionally a 50 caliber machine gun mounted on top of the turret. The tractors were armed with machine guns. In place of a turret it had an open bay

The 747th Amphibious Tank and Tractor Battalion on a training exercise near Point Sal, off the northern coast of Camp Cooke. National Archives.

Soldiers and vehicles of the 747th Amphibious Tank and Tractor Battalion, and the 3623rd Ordnance Company, boarding U.S. Navy LST ships at Point Sal, headed to the Naval Amphibious Training Base at Coronado for three weeks of advanced amphibious training. August 7, 1951. National Archives.

behind the crew, with a dropdown tailgate to carry troops or small vehicles, such as a jeep or a tire-mounted cannon which could be pulled behind a jeep.[22]

While at Cooke, the battalion practiced amphibious landings at Point Sal on the northern edge of Camp Cooke, and live firings at targets on land and at sea. They also occasionally bivouacked on the beach and practiced night maneuvers. The battalion became a cadre unit training replacement soldiers going overseas.[23]

On August 7, 1951, the entire battalion of nearly five hundred men, with all of its vehicles and sixteen men from the 3623rd Ordnance Company, boarded four U.S. Navy LST ships at Point Sal that took them to the Naval Amphibious Training Base at Coronado, near San Diego, California. For the next three weeks the battalion conducted advanced unit training that culminated in a full-scale amphibious assault staged on Coronado's Silver Strand beach, with support from naval landing crafts and Navy jet fighters simulating strafing runs. Nearly 1,000 military personnel and civilians witnessed the main demonstration, which opened with a mock reconnaissance of the landing beach by divers of Underwater Demolition Team 3, who blasted beach obstruc-

Soldiers of the 40th Infantry Division at Mock Village. Originally built during World War II, the village was partially restored for use in the Korean War. February 1951. National Archives.

Troops of 40th Infantry Division going through tactical training problems. February 1951. National Archives.

**The bazooka firing range. Author's collection.**

tions. The battalion returned to Cooke at the conclusion of the exercise on August 21. In December it headed to Fort Irwin in California's Mojave Desert for two weeks of gunnery practice at the Armored Combat Firing Center.[24]

The 40th Infantry Division was still in its early phase of training when Chief of the Army Field Forces Gen. Mark W. Clark visited the 40th to assess its progress on November 16, 1950. Division personnel would have to complete basic and advanced training, and gone through the infiltration, close combat, mock village, and overhead artillery fire courses, before being ready for shipment overseas. On March 28, 1951, four months after Clark's visit, trainloads of 40th Infantry Division troops moved out of Camp Cooke to the San Francisco Port of Embarkation. They were headed to Japan for additional training before arriving in South Korea in February 1952. More than 4,000 soldiers from the Division with less than the fourteen week's training required for overseas shipment remained behind as a rear detachment. After completing the training program, these men would rejoin the 40th in Japan.[25]

The first Air Force personnel at Camp Cooke arrived on January 27, 1951, and established the 3265th Training Squadron, assigned to the 3310th Tech-

**Troops of the 40th Infantry Division during an unidentified military event. Author's collection.**

nical Training Wing at Scott Air Force Base, Illinois. They were organized as a food service unit and attended the Sixth Army's food service sub-school that was established at Cooke two days later as a joint Army and Air Force school to train personnel for careers in food service as cooks and bakers. The training program for the cooks was eight weeks, and three weeks longer for those attending baking courses. Concurrent classes started about every two weeks. By the time the Air Force inactivated the squadron in September 1952, several hundred student airmen (men and women) had graduated from the school.[26]

When Universal Studios began filming the movie *Red Ball Express* about the famed truck convoy of the same name (established in 1944 to keep American frontline troops supplied in Europe), the studio turned to the Army for assistance. In November 1951 the studio arranged with Headquarters Sixth Army to lease sixteen trucks from Camp Cooke's ordnance section. With the studio's own transportation department personnel behind the wheel of each truck, they motored down Highway 101 to Hollywood. The trucks appeared extensively in the 1952 film, which starred Jeff Chandler and Sidney Poitier.[27]

In the 1950s, America was detonating atomic bombs at its Nevada Test Site, more than three-hundred miles east of Camp Cooke. During the first week of February 1951, the *Lompoc Record* reported that city residents and soldiers at Cooke had witnessed distant flashes from atomic explosions on three separate occasions. On September 20, some 1,000 Cooke soldiers headed for the Nevada Test Site, along with several thousand other servicemen from all four armed services, to participate in Operation Buster-Jangle. Conducted between October 22 and November 29, 1951, the operation consisted of six atmospheric nuclear detonations and one underground explosion, collectively known as Desert Rock exercises I, II, and III. The tests were a joint operation between the Department of Defense and the Atomic Energy Commission to psychologically condition soldiers to atomic warfare, and to assess the effects of nuclear detonations on military equipment and fortifications.[28]

The Army units from Camp Cooke taking part in the tests were detachments from the 359th Engineer Battalion, 303rd Signal Service Battalion, 314th Signal Service Battalion, 393rd Ordnance Battalion, 161st Ordnance Depot Company, and the 3623rd Ordnance Company. They were part of various Army units that maintained and operated Camp Desert Rock, an installation of the U.S. Sixth Army south of ground zero. These soldiers provided essential support services while others in the group worked in the forward areas, constructing observation trenches and equipment displays, laying communications, providing transportation, and assisting in other preparations. After the detonations they also recovered test equipment.[29]

Many of the Camp Desert Rock personnel observed at least one test detonation. Before each test they received briefings about nuclear weapons effects and testing procedures. During one test on November 19, the observers left Camp Desert Rock in a bus convoy for the observation location, nine kilometers south of ground zero. When they arrived at the site the men were directed to sit on the ground with their backs toward ground zero. After the initial flash of light from the detonation, they were told to turn and view the fireball and cloud.[30]

Shortly after Cooke sent troops to Nevada, Maj. Gen. William B. Kean, commanding general of the III Corps and director of the Desert Rock exercises, told reporters that the Army was "leaning over backwards" to protect troops from nuclear exposure. Despite assurances from government officials and agencies that nuclear testing was handled safely, public skepticism and questions remain unanswered decades later as to whether hundreds of American servicemen and other citizens were overexposed to radiation, which was responsible for cancers and birth defects.[31]

On March 19, 1951, the hospital at Camp Cooke received its first two battle casualties from Korea. Soon they were coming almost daily, and by the end of the year their numbers had reached 296. An additional 149 soldiers from Korea and 217 from other parts of Asia and Europe with other injuries and diseases were admitted to the hospital in 1951. In 1952 the hospital received 403 military patients from overseas, of which 76 were battle casualties. By the end of the year, Camp Cooke was phasing down, making it unlikely that overseas patients would be sent to the hospital in 1953.

One of the early medical evacuees was nineteen-year-old Pfc. Peter M. Herardo. Semiconscious and with critical wounds to the abdomen, he was airlifted to Cooke on April 20 at the request of his mother. He died six days later, the first war casualty to die at the camp hospital. Herardo was buried in his hometown of Guadalupe, California. Camp commander Col. Frank R. Williams directed that a street at the installation be renamed in honor of the young soldier. The ceremony was held on Saturday morning, May 19, 1951. It was the opening of the first Armed Forces Day celebration at Cooke, a day appropriately set aside by the President of the United States to pay tribute to America's military and a special time to remember its fallen.[32]

Armed Forces Day 1951 turned out to be the year's largest organized event at Camp Cooke. Some 7,000 spectators attended the open house to see military displays and demonstrations presented by Army, Coast Guard, and Hancock College of Aeronautics personnel. Twelve buildings were set up with exhibits. The Army's Chemical Section exhibited special protective suits for use in atomic, biological, and chemical warfare. A display of the latest infantry weapons filled another building. The hospital featured innovative medical devices, including the newest type of iron lung, portable x-ray machines, medical and first aid kits, and other equipment. It also provided demonstrations for treating wounds and snakebites. A fully equipped field ambulance was open for visitors. At a building near the hospital, Army nurses and WACs modeled the latest uniforms. In another area the food service school presented a complete Army field kitchen and served coffee and doughnuts to visitors. The Coast Guard group from Point Arguello displayed a crash boat and rescue equipment. Hancock cadets displayed an aircraft engine and other components. Two theaters were set up to run movies throughout the day, one showing Army combat films from Korea and the other a Red Cross presentation.[33]

Bleachers were rolled out to seat as many visitors as possible for afternoon demonstrations. The show featured a close order drill, crack marksmanship, and the use of various pyrotechnics. The highlight of the show was a tank mounted with a flamethrower destroying a mock pillbox. Army bands and the Santa Ynez high school band furnished music for the festivities.[34]

On June 28, 1951, Governor Earl Warren paid a second visit to Camp Cooke, accompanied by his wife. He observed California Guardsmen from the 49th Infantry Division and the 111th Armored Cavalry Regiment practicing on a firing range. The governor also toured the camp and spoke with groups of Guardsmen. He proudly awarded the California Guard commendation ribbon to thirteen officers and enlisted men of the 49th Infantry Division for their service in organizing, administering, and instructing the first National Guard officer candidate school in the state.[35]

From June to August 1951, no less than a dozen California Reserve and Guard units conducted two weeks of field training at Cooke. They included 136 personnel from the 1st Division, California National Guard Reserve; 125 officers and enlisted men from the 2nd Division, California National Guard Reserve; 200 reservists from the 325th Ordnance Reserve Group; 8,000 Guardsmen from the 49th Infantry Division; 500 Guardsmen from the 111th Armored Cavalry Regiment; 2,000 reservists from the 91st Infantry Division; an undisclosed number from the 349th General Hospital, a reserve unit from the Los Angeles area; 150 reservists from the 823rd Medical Station Hospital and the 394th Field Artillery Battalion; an unreported number of men from the 499th Engineer Aviation Battalion and the 362nd Engineers Group; and 3,000 reservists from the 13th Armored Reserve Division, including attached units. The 362nd Engineers was the lead organization that

The 466th Antiaircraft Artillery Battalion arriving at Camp Cooke. January 15, 1952. Author's collection.

**The first men of the 44th Infantry Division to arrive at Camp Cooke were soldiers of the 123rd Infantry Regiment from Springfield, Illinois. After detraining at the rail-head, the troops boarded buses that took them to their barracks. February 21, 1952. National Archives.**

dedicated part of its training time to construct a 2,500-foot-long extension to the camp's airstrip.[36]

In July the *Cooke Clarion* announced the opening of a golf driving range opposite the camp hospital. This was almost certainly a renovation of the nine-hole course established by the Army in July 1945. Capt. Charles E. Hughes of the hospital medics designed the original course. It featured a 35 par course of 2,887 yards, with two par three holes, six par four holes, and one 500-yard par five hole. The course covered the area in front of and on the side of the hospital. Although long removed, on current maps of Vandenberg AFB the course would be bordered by Herardo, Nebraska, South Dakota, and Ocean View Avenues.[37]

In the fall of 1951 a critical shortage of agricultural laborers in the Santa Maria and Lompoc valleys threatened to disrupt the tomato harvest. When the same problem emerged in 1943, farmers appealed to the Army for assistance and received more than enough labor to save the crop. For the current emergency the Army again allowed soldiers to harvest the crop during their off-duty hours. In October more than four-hundred soldiers from Camp Cooke picked tomatoes in the Lompoc, Santa Maria, and Santa Ynez Valleys. Each soldier was paid for his labor, depending upon the total number of tomatoes picked.[38]

On November 19, 1951, Col. Alexander G. Kirby succeeded Col. Frank

**Illinois Governor Adlai Stevenson is escorted by an unidentified Army officer, and followed by Lt. Gen. Joseph M. Swing (left) and Maj. Gen. Harry L. Bolen, shortly after his arrival at Camp Cooke. May 7, 1952. Author's collection.**

R. Williams as the second post commander at Camp Cooke since its reactivation in 1950.[39]

The year 1952 brought new Army units to Camp Cooke, beginning with the 466th Antiaircraft Artillery Automatic Weapons Battalion (Semi-Mobile) on January 16. Led by Lt. Col. Louis Eisen, the 466th was an Organized Reserve Corps unit from Virginia and one of the few remaining all-black organizations in the Army at that time. Most of its officers were white. It arrived at Cooke after several months of training at Camp Edwards, Massachusetts. When it arrived at Cooke, the 466th was reassigned from the First Army to the 250th Group, Western Army Antiaircraft Command.[40]

Less than a month later, on February 13, the first troops of the 44th Infantry Division, an Illinois National Guard unit called into federal service, detrained at Camp Cooke. Led by Brig. Gen. Paul K. MacDonald, the advanced detachment brought with them a trainload of heavy equipment, ordnance supplies, and vehicles. The main body of 4,000 division troops arrived on February 19–23. Greeting them as they arrived was their division commander, Maj. Gen. Harry L. Bolen, and Col. Kirby, post commander. Some Guardsmen elected to travel to Cooke by personal automobile with their families. They quickly discovered the severe housing shortage that forced many families to settle for whatever they could find and sometimes at excessive rents. To help bring the division up to combat strength, over the next two months the Army rotated into the unit several thousand men returning from Korea and moved in additional Stateside troops from Camp Roberts, California; Fort Ord, California; Camp Chaffee, Arkansas; and Fort Leonard Wood, Missouri.[41]

Visiting Cooke on April 22–23, 1952, Chief of the Army Field Forces, Gen. Mark W. Clark, told the 44th that its mission had changed from a combat readiness division to a cadre unit, training and preparing soldiers for combat as individual replacements for overseas units. Throughout the remainder of the year the division transferred out several thousand trained soldiers to Europe and Korea, while continually receiving Korean War returnees who greatly aided in the training of levied enlistees and inductees.[42]

On May 7, 1952, Illinois governor Adlai Stevenson flew into Camp Cooke for two days of meetings and tours. Accompanying the governor was a military aid and Sixth Army commanding officer Lt. Gen. Joseph M. Swing. The governor met with men of the 44th Infantry Division from his home state, lunched with troops of Company I, 129th Infantry Regiment, had a personal visit with the division commander, Maj. Gen. Bolen, conducted a press conference, inspected training facilities, and observed a division review of 15,000 soldiers. He also addressed a group of thirty-eight Santa Maria Union High School students from the Junior Statesmen chapter and answered their questions. The Army bussed the students between their school and Camp Cooke.[43]

May 17, 1952, marked the second annual Armed Forces Day celebration at Camp Cooke with an open house invitation for the public to come out and enjoy the festivities. The climax of the day came during the "mad minute" at 1:30 p.m. when every available weapon, from machine guns to mortars and tank-mounted 75mm field guns, all went off at once to demonstrate the concentrated firepower of an infantry unit. Spectators seated in bleachers also witnessed individual weapons firing demonstrations. A rifle squad showed how it advances upon and destroys an enemy pillbox position with the aid of flamethrowers, rocket launchers, and live explosive charges. Most of the equipment demonstrated was available for close inspection by the public. Among the other exhibits were armored vehicles, an Army helicopter, hospital equipment, and a complete field kitchen. Visitors also enjoyed guided tours of the post and jeep rides. To make the most of the day, visitors were encouraged to bring picnic lunches, or, if they preferred, they could buy a meal prepared in Army kitchens.[44]

Illiteracy among some soldiers at Camp Cooke was a problem the Army attempted to correct by having every man earn a high school diploma or the equivalent. Accomplishment of this goal would improve unit efficiencies and greatly benefit the men upon their return to civilian life. At the 747th Amphibious Tank and Tractor Battalion, every man eligible was encouraged to participate in the program. Those who were interested and close to graduation from high school before entering the Army were given General Educational Development tests. If they passed the tests they would either receive a diploma from the school they left, or, in cases where schools did not recognize the results of the Army test, the State in which the men resided would issue a certificate equivalent to a high school diploma. Those who were a few years away from twelfth grade would be given time and materials to study and prepare for the test.[45]

All servicemen with a minimal education were encouraged to attend a special primary school for grades one through five that was established under the supervision of the Post Information and Education Officer. Staffed by trained teachers, the school consisted of five classrooms and was in continual operation. To help ensure the program's success, Headquarters Sixth Army directed that men attending the 12-week course be given four hours of duty time a day for school, and be relieved of company responsibilities that might interfere with their education.[46]

A medium tank undergoing repairs at the Post Ordnance Shop in building 11352 is shown to guests from Santa Maria by Capt. Thomas Wilson, 393rd Ordnance and Maintenance officer, during a tour of the post and the 44th Infantry Division on October 9, 1952. This building is one of the few structures from Camp Cooke that remain standing at present day Vandenberg Air Force Base. U.S. Army/author's collection.

Each candidate completed an aptitude test to determine his appropriate grade level. Should the candidate need to go into the first grade, he would begin with learning phonetics and progress through the standard curriculum of reading, writing, arithmetic, and government. At the end of twelve weeks, if the GI failed to pass the course his case would come up for review. He might be allowed to repeat the course, be discharged on grounds of a mental handicap, or sent to a labor battalion if he had the ability but refused to learn.[47]

From September 1950, when the program began, through the middle of August 1952 some 2,000 men graduated from the primary school course. Many of the graduates went on to earn high school diplomas by enrolling in correspondence courses available to all service personnel through the United States Armed Forces Institute (USAFI).[48]

Beginning on April 3, 1951, men at the college level could register for extension courses taught at Cooke by professors from the University of California at Santa Barbara. Initially, three classes were offered—in political science, English, and criminology. The USAFI subsidized the regular class fee of $27, leaving a balance of $6.75 to be paid by the enrollee.[49]

Between July and August 1952, multiple Army Reserve units descended on Camp Cooke for two weeks of required field training. They included the 347th General Hospital from Palo Alto, California, and nineteen units attached to the 311th Logistics Command and the 349th Transportation Traffic Regulation Group from Southern California and Arizona. Also at Cooke were two psychological warfare units, the 306th Radio Broadcasting and Leaflet Company from the Greater Los Angeles area, whose mission was to develop and drop propaganda leaflets called T-[truth] bombs on enemy homelands and behind frontline areas, and the 352nd Loud Speaker and Leaflet Company from Seattle, Washington. The latter unit would operate close to the front lines.[50]

One of the palpable effects of the Korean War, and President Truman's hard-line policy against Communism, was a huge military build-up. Defense spending, as a percentage of the gross domestic product, almost tripled between 1950 and 1953 (from 5 to 14.2 percent). With so many tax dollars going to support the armed services, election-year politics of 1952 picked up on the theme of wasteful military spending. At Camp Cooke the Army reacted to the criticism by inviting nineteen community representatives from Santa Maria for a tour of the camp and the 44th Infantry Division. Those attending the open house on October 9, 1952, were Glen Seaman, mayor of Santa Maria; Robert Seavers, Chamber of Commerce; York Peterson, city engineer; Frank Crakes, fire chief; Frank McCaslin, chief of police; John Groom, KSMA radio station; William Cook, KCOY television; Carroll Gerwin, *Santa Maria*

*Times*; John Ball, *Santa Barbara News-Press*; B. A. Devine, manager of Southern Counties Gas Company; Frank Brown, Rotary Club; Ralph Sutton, Lions Club; and Joseph Sesto, insurance sales. No reports are available about their impressions of the camp, but judging from Army photographs of the group they seemed to have enjoyed their visit.[51]

# 5

## ORGANIZED ENTERTAINMENT AT CAMP COOKE

### Entertainment During World War II

Organized civilian entertainment for soldiers at Camp Cooke was non-existent inside the camp during the first five months following the camp's activation in October 1941. The few diversions available to soldiers outside the camp were at least ten miles away in the community of Lompoc, and more than twenty miles to Santa Maria. Initially, the selection in these towns was confined to a couple of movie houses, churches, a few questionable establishments, and not much else. The first in a series of dances for servicemen, under the sponsorship of the United Service Organizations (USO), in Lompoc was held at the high school gym on January 30, 1942. Soon after, fraternity groups began sponsoring dances in their halls. Another Lompoc group was the Home Hospitality Committee of the Defense Recreation Committee. Chaired by a clergyman, the committee invited soldiers to attend church services with its members, who represented different denominations. Following services, they and other congregants would take two or more soldiers to their homes for dinner.[1]

Because Camp Cooke was far from large metropolitan centers, it was both a blessing and a problem for the Army. Away from big cities, soldiers were less likely to stray into undesirable activities. But an isolated environment without healthy alternatives on which soldiers could spend their free time could breed other disagreeable problems. To maintain high morale and good order among its troops, the Army designed Camp Cooke with two service clubs that included libraries, a Sports Arena, four theaters, four gyms (called Field Houses), dozens of ballparks, and three hobby shops as outlets for entertainment and recreation.[2]

Shortly after the 5th Armored Division arrived at Camp Cooke, the first two theaters opened at the camp in early March 1942, followed by Service Club No. 1 in April and the Sports Arena a month later. USO clubs soon appeared in Lompoc, Santa Maria, and at Surf. In early 1943, two more theaters and a second service club also opened at Cooke.[3]

March 3, 1942, marked the start of organized entertainment in the camp when comedian Bob Hope became the first major celebrity entertainer to visit. He transported his nationwide radio program from Los Angeles to Camp Cooke to perform before an exuberant audience of soldiers in Theater No. 1. His broadcast officially opened the newly built show house on California Boulevard between 25th and 26th Streets.[4]

The show featured Frances Langford, Jerry Colonna, Betty Hutton, Skinnay Ennis, and special guest Yankee baseball great Babe Ruth. Larry Keating of NBC radio was the announcer. The "Six Shots and a Miss," a reference to the six male musicians and one female member of the orchestra, covered the musical portion of the show. Following the introduction, Hope opened the broadcast with his usual snappy monologue, poking fun at army life. He then chatted with the cast and guest star, and offered a closing skit with Colonna. In separate sections between the comedy, Langford sang "I've Got It Bad (and That Ain't Good)," and Hutton sang a Dixieland band selection. During two commercial breaks, Keating advertised the show's product sponsor, Pepsodent toothpaste, and reminded listeners to buy war bonds and stamps. Hope closed the show with an adaptation of his signature theme song, followed by an appeal for bond sales and support for the Salvation Army:

> Thanks for the memory
> You fellows of Camp Cooke;
> all wear that victory look...[5]

The next day the second of four theaters, with up to 1,200 seats, opened at Cooke with a musical performance by Benny Meroff and his orchestra, best known for their 1928 recording "Smiling Skies." The theater was on California Boulevard between 12th and 13th streets.[6]

Clowning aside, Harpo Marx played wonderful music on the harp and, with the flamboyant Ada Leonard and her All-American Girl orchestra, gave two evening performances in Theater No. 1 on March 16, 1942. In the orchestra were Jane Sager (one of the world's best trumpet players), Edith Lawrence, Brownie Slade, Midge Goodrich, Joann Koupis, Mildred Cobb, Connie Van, Bernice Little, Dez Thompson, Clara Friend, and Ethel Button. The troupe also included the Three Sophisticated Ladies, a dancing, singing, knockabout comedy act; the Blossom Sisters (Helen and Dorothy), who featured comedy dance routines; and Shirley Lloyd, actress and singer.[7]

**Left to right: Skinnay Ennis, Francis Langford, Bob Hope, post commander Lt. Col. Carle H. Belt, Jerry Colonna, Betty Hutton, and Babe Ruth at Theater No. 1. March 3, 1942. U.S. Army/*Cooke Clarion*.**

The opening of the first of two service clubs at Cooke on April 11, 1942, gave soldiers and civilians the opportunity to meet new friends, to dance, relax, and enjoy a meal in a modern facility. To mark the occasion, more than 250 chaperoned young ladies from the neighboring towns were escorted in Army trucks and buses to participate in a gala celebration as dance partners for soldiers.[8]

Service Club No. 1 was a two-story, 15,431 square foot building across from Division Headquarters. Its center auditorium could accommodate three-hundred couples at a dance and more than 1,000 guests at a program. Open from 8:00 a.m. to 11:00 p.m. seven days a week, the club offered continuous nightly entertainment, with special variety shows at least once a week. Located in the balcony above the main auditorium was a library with some 10,000 books available for loan.[9]

Musically, the 5th Armored Division furnished the club a 60-piece orchestra, two regimental bands, and two dance bands. After the division shipped out in 1943, similar musical arrangements were made with other Army units at Cooke.[10]

The service club was operated by the Special Services Office of Corps Area Service Command 1908 (later Service Command Unit 1908), which was headed by an Army major with a staff of twenty enlisted personnel. Daily

Service Club No. 1 at Camp Cooke. Courtesy the Lompoc Valley Historical Society.

The main room at Service Club No. 1. The wonderful art deco mural on the wall records some of the activities at the camp. During dances or other special events the chairs on the first floor would be placed along the walls. U.S. Army booklet *A Camera Trip Through Camp Cooke*.

operations at the club were handled by a staff of forty civilian employees that included a principle hostess, at least two junior hostesses, and a librarian.[11]

Service Club No. 1 was the central exchange for all Army organizations at Cooke. At its peak operation in 1943, its subdivisions included a second service club, two guest houses, four theaters, a 4,000-seat Sports Arena (later

**The Sports Arena at building 9008. February 23, 1942. U.S. Army/author's collection.**

expanded to 5,000 seats), and the weekly *Cooke Clarion* newspaper. The 24,667 square foot Sports Arena on Washington Avenue near California Boulevard opened to a musical variety show on May 7, 1942. Because of its size, the Arena attracted productions that could draw large audiences to fill its cavernous auditorium.[12]

For Army brass, an Officers' Club opened in building 6001 behind Division Headquarters on May 23, 1942. Among the special guests attending the opening dance party were Maj. Gen. Jack W. Heard, 5th Armored Division commander; Col. Carle H. Belt, post commander; Charles L. Preisker, chairman of the Santa Barbara County board of supervisors; Marion B. Rice, mayor of Santa Maria; and their respective wives. A second club, or Post Officers' Mess, opened a year later in building 16032 on G Street between Montana and Wyoming Avenues.[13]

On May 25–26, 1942, USO–Camp Shows, Inc., presented the comedy variety show *Hollywood Follies* at the Sports Arena in two evening performances. Heading the list of entertainers were the Three Stooges. The three-man wrecking crew who typically beat each other senseless and were apparently impervious to physical injury, were donating their services during their two weeks of play with the troupe. The other thirty-nine cast members included the Fanchonettes, a precision line dance of sixteen pulchritudinous young women; Frank Gaby, comedian and ventriloquist; Rae and the Rudells, a comical trampoline trio of two men and a woman; the Hackers, satirical ballroom dancers; Betty Walters, contortionist; Evers (Frank) and Dolores, tight wire walking duo; and Dick, Don, and Dinah, knockabout acrobatics.[14]

*What's Cookin',* which played at the Sports Arena on June 11, was an original musical comedy revue put together by talented artists from the 5th

Armored Division. Privates Sid Tepper and Phil Weiner wrote many of the songs for the show. Weiner was a former arranger for singer Vaughn Monroe and Barney Rapp's Orchestra. Tepper was a staff composer at Broadcast Music, Inc., and later went on to write songs for Elvis Presley and other artists.[15]

A few days later, on June 15, the cast of the Santa Barbara Civic Theater presented at Theater No. 1 *Room Service*, the successful Broadway comedy later made into a movie.[16]

A third movie theater opened on Utah Avenue on June 21 with the showing of *Private Buckaroo*. The small, 256-seat theater was replaced by a much larger building in February 1943.[17]

Coming from the world-famous Roxy Theater in New York City, USO–Camp Shows, Inc., presented the *Roxy Revue* at the Sports Arena in two performances on June 22–23, 1942. The hit musical with thirty-three performers included a dance chorus line of sixteen attractive Gae Foster show girls, comedian and mandolin player Dave Appollon, blues radio singer Barbara Lamarr, celebrity impersonators the Wesson Brothers (Dick and Gene), rhythm and tap dancer Eleanor Teeman, comic juggler Stan Kavanagh, comic acrobats Marion Belett and the English Brothers, and tap dancer Linda Moody.[18]

On June 28, 1942, an estimated 15,000 khaki-clad soldiers from the 5th Armored Division and civilian guests cheered wildly at a Wild West rodeo show at Camp Cooke organized by Santa Barbara County cattlemen and division personnel, with a special guest appearance by actor Leo Carrillo. The rodeo featured a comic routine in which a few division soldiers drove Army jeeps, while, standing in the back of each vehicle, professional cowboys from the county flipped their lariats over the horns of fleeing, bewildered steers. Other attractions from the nearby Santa Ynez Valley in Santa Barbara County were Dwight Murphy's famous string of golden palomino horses from Rancho San Fernando Rey, and Lucius B. Manning's Tennessee walking horses from Rancho Piocha. The show opened with a parade of tanks and other mechanized vehicles.[19]

USO service clubs became surrogate homes for many lonely GIs separated from family and friends, some for the first time. Here they could dance, join in a songfest, play board games, participate in a show or be a spectator at performances, record a phonograph record to send home, or just relax while munching on a snack. The clubs provided countless other services, including a place to shower or shave, postal services, and even sewing on a new chevron. They also assisted the wives of servicemen in finding a room or apartment. When housing could not be found, the clubs provided temporary accommodations at a USO dormitory.[20]

Women made up the core of volunteers at all USO clubs and served as senior and junior hostesses. The latter had to be at least 18 years of age. Their job was to keep up morale by bringing a little sunshine into the soldier's life. Usually that meant a smile and providing a friendly or sympathetic ear to homesick soldiers. Hostesses were held to strict standards of conduct to maintain the USO's clean reputation. Still, friendships between volunteers and soldiers sometimes led to romance and on more than one occasion blossomed into marriages.[21]

In Santa Maria, the first USO club opened on March 1, 1942, to a crowd of soldiers from Cooke and aviation personnel from the nearby Hancock College of Aeronautics. A formal dedication ceremony occurred two weeks later on March 15. The clubhouse was established in the city's old public library building, known as the Carnegie Library, at 410 South Broadway. Before opening the building to soldiers, its interior was refurbished and redecorated by a committee of volunteers, headed by J. Ben Wiley and Mayor Marion B. Rice. The club operated under the auspices of the YMCA, with Huge E. Parminter as its director, and Dan Kaplan of the Jewish Welfare Board as assistant director.[22]

A year later, on February 12, 1943, the city of Santa Maria opened a USO dormitory at 208 North Broadway, formerly used as the Eagles Hall. For soldiers making the most of a weekend pass to the city, this was a welcomed refuge.[23]

**The USO Club (right) on South H Street in Lompoc. Courtesy the Lompoc Valley Historical Society.**

Unlike the armed forces that was racially segregated during World War II, the USO was opposed to racial segregation but chose to avoid conflict with local white communities if they opposed integration. This never became an issue in Lompoc because no black soldiers were stationed at Camp Cooke. For black servicemen stationed at the Santa Maria Army Airfield, a separate USO club operated at 511 West Main Street, and eventually closed at the end of November 1945.[24]

The first USO club in Lompoc opened in the Foresters' Hall, formerly known as the Dinwiddie building, at 116 South H Street. Several hundred soldiers from Camp Cooke and a near equal number of Lompoc residents attended the opening dedication ceremonies on July 11, 1942. Following a few brief remarks and a short musical serenade from an Army regimental band, a

A group of Junior Hostesses signing registration cards at the table where Senior Hostesses Mrs. J. H. Merriam and Mrs. Alden Lewis are seated, before going into a dance at the Walnut Avenue USO building. Until permanent admittance cards were issued to the Junior Hostesses, every girl wishing to attend the USO dance had to fill out a card giving certain information about herself and the names of two local residents as references. October 1942. Courtesy the *Lompoc Record.*

color guard from the 5th Armored Division conducted the flag raising ceremony, signaling the beginning of club operations. The club was open daily from 11:00 a.m. to 11:00 p.m. and was under the direction of Edward J. Trainor and assistant director Joseph P. Hiss of the National Catholic Community Service organization.[25]

On August 7, 1942, a second USO club opened in Lompoc, at 119 West Walnut Avenue, and was dubbed the Walnut Street club. It was operated jointly by the Salvation Army, with Adjutant Cecil L. Day as its first director and Mrs. Bayonne Glenn of the YWCA as associate director. The hundreds of townspeople who attended the opening day ceremonies and open house in the evening heard speakers from the federal government, the mayor of Lompoc, the USO, the Salvation Army, the YWCA, Camp Cooke, and the Coast Guard. The new, two-story building was constructed by the C. W. Driver Company of Los Angeles under a 60-day construction contract funded by the Federal Works Authority for $98,000. The main floor of the building included a snack bar, offices for the USO staff, a stage, and dance floor. The upstairs "penthouse" was used for meetings and included a fully equipped kitchen.[26]

The YWCA established a training program for hostesses at the Walnut Avenue USO club that followed closely the national USO policy rules, and probably a few of their own. Hostesses were advised to make themselves attractive but not glamorous, to be interested in something besides themselves, to be able to dance well, play ping pong adequately, and to know how to make men feel at home. They also had to provide the names of two local residents willing to vouch for their moral character. Hostesses were not permitted to wear slacks. Girls wanting to attend dances had to have a permanent registration card, or fill out a card at the door, which included the two personal references.[27]

At Camp Cooke, film star Ilona Massey was the lead attraction in USO–Camp Shows' *Full Speed Ahead*, performed before two full houses at the Sports Arena on July 20–21, 1942. Comedian Don Rice emceed the show, which included a chorus line of sixteen glamorous ladies from the New York City stage, juggling sensation Ben Beri, singer Ginger Harmon, tap dancer Shirley Van, wire-walker Harold Barnes, and comedic contortionists Jimmy Shea and Gene Raymond.[28]

USO–Camp Shows' next presentation at the Sports Arena was *Crazy Show* on August 24. A deliriously funny Broadway musical revue with twenty-four headline performers, it starred Milt Britton and his orchestra. Described as one of the finest entertainment bands assembled, they were noted for zany, prop-smashing, food-fight, acrobatic antics. Also on stage were the Kim Loo Sisters (Patricia, Margaret, and Alice), popular swing singers; Frank Ross and

Anita La Pierre, comedy impersonators; the Honey Family, three men and three women presenting fast-paced dancing with acrobatics; and comedian-musician Ted Lester, who played regular and miniature musical instruments, as well as air pumps and balloons to get an unusual mixture of acoustics.[29]

On October 15, 1942, Service Club No. 1 was transformed into one of the great music halls of the world when world-renowned violinist Jascha Heifetz and acclaimed classical pianist Emanuel Bay staged a concert performance for 1,000 soldiers and their guests. The show was arranged by the Camp Special Services Office and the Hollywood Victory Committee, Inc.[30]

Shortly after the United States entered the war, the R. J. Reynolds Tobacco Company, makers of Camel cigarettes, sent its musical variety radio program, *The Camel Caravan*, on tour to military installations around the country. While the company may have been flushed with patriotic spirit to sponsor these traveling shows, it made good use of them to promote its product. At the end of each show, shapely young girls—dubbed Camelettes and dressed like drum majorettes, with short skirts and white boots—passed through the audiences handing out free cigarettes.[31]

On October 27, 1942, the *Caravan* presented two variety shows at Cooke, an afternoon performance at the Station Hospital and an evening performance at the Sports Arena. The cast included entertainer Joey Rardin as emcee; Fid Gordon, comedy violinist; the Texas Rangers, a seven-man combination western musical group that had their own radio show and appeared in movies; Paul and Paulette (George and Paulette Paul), a husband and wife trampoline act; and Darlene Ottum, a tap and acrobatic dancer.[32]

The music and wacky antics of *Kay Kyser's Kollege of Musical Knowledge* program played to a nationwide radio audience and to a packed house at the Sport Arena on December 2, 1942. The two one-hour shows starred bandleader Kay Kyser and the entire orchestra, Ish Kabibble (Merwyn Bogue), Julie Conrad, and announcer John "Bud" Hiestand. The show featured songs and comedy routines wrapped around musical questions asked by Kyser dressed as the "old professor." First and second prizewinners from the audience received $50 and $25 war bonds, respectively. Other contestants won $10 in war savings stamps and cartons of Lucky Strike cigarettes, the sponsor of the show.[33]

On December 5 and 7 the Sports Arena again rocked with laughter as USO–Camp Shows Inc. presented the musical comedy *Hollywood on Parade*, with singing comedian Bert Frohman as emcee. Playing before a large and enthusiastic audience, the glamour town revue featured a chorus line of sixteen girls; magician Cardini and his wife performing sleight-of-hand card tricks; the singing Tanner Sisters; the acrobatic dance act of Joe and Jane McKenna; and two dance teams, Allen and Kent, and Martez and Dolita.[34]

The year 1942 closed out with one of Broadway's successful hit comedies, *Arsenic and Old Lace*, performing a two-night run on December 19 and 21 at Theater No. 1. This was the first USO–Camp Shows play to be staged at Cooke. The cast was composed of veteran actors from the New York stage.[35]

Bert Frohman at the microphone with the chorus line on stage in USO–Camp Shows' *Hollywood on Parade.* U.S. Army/*Cooke Clarion.*

Playing before two highly appreciative audiences, the USO–Camp Shows' Broadway musical comedy *Flying Colors* scored big at the Sports Arena on January 2 and 4, 1943. The lavish show featured a variety revue of eight vaudeville acts of comedy, song, dance, and acrobatic talent. Included in the cast were Lew Hearn, Johnny Masters, Rowena Rollins, the comedy of the Arnaut Brothers, singer Barbara Lamarr, vocalist Mark Plant, the three Winter Sisters in a high-spirited tap and acrobatic repertoire, and a chorus line of twelve Gae Foster starlets singing and dancing. The girls closed the show with a difficult routine on roller skates.[36]

Making her second appearance at Cooke in less than a year with her own show, Ada Leonard and her All-American Girl orchestra received standing ovations from 10,000 soldiers and civilians in two performances at the Sports Arena on January 18–19, 1943. *Yank* magazine reported that Leonard received her loudest applause when she experienced what we would now call a "wardrobe malfunction." Her skirt fell down during a dance number, and the soldier audience thought she was going to do a strip tease. She quickly snapped the skirt back into place and finished the dance. The cheering audience became silent.[37]

Also in the troupe were distaff vocalists Martha Stewart and Elinore Sherry in separate acts; the comedy act of Lynn Russell and Marion (or Miriam) Farrar; a tap dancing routine by Key (or Kay) Taylor; and puppet master Catherine Westfield with her near life-size puppets that included Donald Duck on skates, and caricatures of movie celebrities Caesar Romero and Veronica Lake. A special afternoon performance was also held at the Red Cross recreational hall for hospital patients and staff. In this and at other hospital shows, entertainers also met with ward patients individually and in small groups.[38]

On the evening of January 28, 1943, the Hancock Music Ensemble thrilled music lovers at Camp Cooke with a melodious mixture of classical and popular tunes at Service Club No. 1. Joining the ensemble for his first performance at Cooke was the baritone voice of John Raitt. A future star of musical theater and stage, Raitt would later become the father of Bonnie Raitt, who achieved musical acclaim in her own right thirty years later. The ensemble came to Camp Cooke through the courtesy of the Hancock Foundation for Scientific Research at the University of Southern California. It was founded by Captain G. Allan Hancock, owner-operator of the Hancock College of Aeronautics in Santa Maria, and cellist for the group. Hancock was a businessman, aviator, explorer, educator, railroad engineer, and philanthropist.[39]

Two days later, on January 30, Sports Shows, Inc., presented several boxing and wrestling exhibitions at Cooke's Sports Arena. The featured attractions for the night were appearances by world boxing champions Jim Jeffries, Willie Richie, Jack Root, and Ceferino Garcia.[40]

The Broadway comedy bombshell *Junior Miss*, the second play presented at Cooke by USO–Camp Shows Inc., held two evening performances at Theater No. 1 on February 3–4. The cast of seventeen players included Lucille Fetherston and Strelsa Leeds.[41]

February 13, 1943, brought to the Sports Arena General Electric's "House of Magic." A hit at the San Francisco and New York World's Fairs of 1939, the show presented amazing scientific technologies for that era. Among the demonstrations were a model train that obeyed verbal commands, music sent across the room on a beam of light, and an electric lamp lit with a match.[42]

That same month, on the 24th, Warner Brothers Studios began filming at Camp Cooke its version of Irving Berlin's stage hit *This Is the Army*. Shot in Technicolor, the movie featured actors George Murphy, Ronald Reagan, Joan Leslie, George Tobias, and Alan Hale, with appearances by vocalist Frances Langford and the powerful contralto voice of Kate Smith. Mingling among some three-hundred regular soldiers from the camp employed as extras for various training scenes, the soldier-actors sang, danced, and marched. World heavyweight boxing champion Joe Louis, now a sergeant in the Army, appeared as himself in an outdoor boxing exhibition. Louis had been assigned to a duty station near Hollywood, and for the movie was temporarily assigned to the 5th Armored Division.[43]

Broadway and Hollywood screen actress Grace McDonald took center stage on February 27, 1943, as the star of *In the Groove*, a USO–Camp Shows musical revue held at the Sports Arena. Also in the show were the high-swinging band of Bill Bardo, the Del Rios acrobatic team, celebrity impersonator Wally West, vocalist Judy Powers, vaudeville comedians Lewis and Ames, and the six dancing Tip Top Girls. An encore performance was held three days later.[44]

The new Theater No. 3, on Utah Avenue between Colorado and South Dakota Avenues, opened on the evening of February 26, 1943, with the showing of *Random Harvest*, starring Greer Garson and Ronald Coleman. The building possessed the latest in sound and projection equipment, and the modern marvel of air conditioning.

Making his appearance at the theater on March 2, the peripatetic Bob Hope filled the house with GIs eager to hear and watch the nationwide broadcast of his weekly radio program. Featured in this week's star-laden cast were Jerry Colonna, Frances Langford, Barbara Jo Allen (playing the bit character Vera Vague), Maria Montez, announcer Wendell Niles, and Skinnay Ennis and his orchestra (with vocalist and guitarist Tony Romano). Immediately following this 30-minute funfest of gags, wisecracks, and songs, Hope moved the entire show to the Sports Arena, where more than 5,000 soldiers greeted the cast with thunderous applause.[45]

Theater No. 4, on G Street between Montana and Wyoming Avenues, opened on March 6, 1943, with the feature movie *Forever and a Day*. This theater was identical in design to the new Theater No. 3.[46]

On March 7 the second service club opened in building 16027 on Utah Avenue between Montana and Wyoming Avenues. Although no special events were scheduled on its first night, in the days that followed the club served the camp population with additional recreational and entertainment activities, including dances, talent shows, hobby classes and the like. The club had a cafeteria, a soda fountain, a lounge, and a library with more than 7,000 books.[47]

*Claudia*, a new comedy hit from Broadway, made its way to Camp Cooke on March 15, 1943. It was sponsored by USO–Camp Shows. Emily McNair, Edward Harvey, and Myrtle Tannahill filled the leading roles. The comedy played Theater No. 1 the first night and moved to Theater No. 3 on the second night.[48]

Beginning on March 21, 1943, radio station KTMS in Santa Barbara started broadcasting a weekly Sunday evening show from Service Club No. 1. These series of half-hour shows were produced by the Post Special Services staff at SCU 1908. They showcased live music from different regimental bands, original comedy skits, stories, and camp news. The broadcasts continued for about a year.[49]

Even the Czar of Russia would have been pleased—if he were alive—with the musical repertoire of the world famous General Platoff Don Cossack Chorus. Under the direction of Nikolai Kostriukov, the chorus presented a lively selection of Russian folk, operatic, and military songs in two evening shows at Theater No. 3 on April 2–3, 1943. The former Cossack cavalry officers of the Russian Imperial Army integrated their singing with spirited traditional Cossack dancing, most notably, the famous "Cozatzka" knife twirling dance. USO–Camp Shows sponsored the event.[50]

In April 1943, 272 men of the 6th Armored Division participated in the production of *Baptism of Fire*, one in a series of U.S. Army Ground Forces training films collectively called "Fighting Men." Traveling to the Warner Brothers ranch in Southern California, the military detachments comprised the majority of the personnel in the picture.[51]

The 35-minute film tells the story of an average private as he goes through his first ordeal in battle and comes out a "good soldier" ready to carry on with the fight. The picture deals with battle anxiety. With several graphic fighting scenes, it was designed to prepare the soldier for combat with the enemy. William "Breezy" Reeves Eason of Warner Brothers Studios directed the film, with Maj. Joseph W. Langston as technical advisor. The film was nominated for an Academy Award for Best Documentary in 1944. While bivouacked at the ranch, the men also took part in the production of *Secret Weapons*, another in the "Fighting Men" series.[52]

On April 9 the Hancock Music Ensemble, noted radio and concert stage musicians from the University of Southern California, played another of its many fine concert series at Cooke, with John Raitt, operatic vocalist, returning for a second guest performance as the featured soloist. The performance was at Service Club No. 2.[53]

USO–Camp Shows presented John Olsen and Harold Johnson's madcap comedy hit *Hellzapoppin* at the Sports Arena on April 20–21. Originally produced as a play in 1938 and three years later turned into a movie, this current production, with some forty-three players, continued the gags, absurdities, and improbable situations that made the show a success. Leading cast members included entertainer Milton Douglas, singer Jack Leonard, juggler Ben Beri, the comedy of the Emerald Sisters, and the high kicks of Dorothy Deering and a chorus line of sixteen Roxyette dancers beautifully costumed.[54]

The next USO–Camp Show event, on May 24–25, was *The Band Wagon*, a variety show with the Gray Gordon Orchestra, the two Whitson Brothers acrobatic team, songstress Del Parker, comedian Willie Shore, the Risley comedy acrobatic trio, dancer Eve Matthews and various other acts.[55]

Top-ranking stage favorites from vaudeville and Broadway comprised the return of the *Roxy Revue* to Camp Cooke exactly one year after it played at the Sports Arena. USO–Camp Shows presented the second showing, with a new cast, on June 8–9, 1943. It featured a chorus line of sixteen Gae Foster girls; comedy musical impressionist Bert Lynn; juggler James Evans; the Three

Reddingtons in a humorous trampoline act; comedian Charlie Kemper, whose skit poking fun at First Sergeants resonated with the men; and the delightful vocals of the Read Sisters, Floy and Martha.[56]

Baseball, whether as a spectator or participant, was another well-attended pastime by soldiers in the camp. On June 12 and 13, 1943, in a doubleheader played at Camp Cooke between competing Army teams, the Post All-Stars and the 6th Armored Division took on the Santa Ana Flyers of the Santa Ana Army Air Base. Leading the Flyers to a 2–0 and 2–1 victory over the Cooke teams was former Yankee slugger Cpl. Joe DiMaggio, who pounded out one hit on Saturday's game and two in Sunday's tussle.[57]

In honor of Father's Day on June 20, Camp Cooke extended an invitation to all dads from Santa Barbara to Santa Maria to spend the day with soldiers at the camp. The first visitors began arriving at the camp around noon and were directed to key locations where groups of soldiers in Army vehicles drove them to various mess halls for lunch and then to the baseball field across from the Sports Arena. There, some 4,000 spectators sat on bleachers and on the ground to watch demonstrations of mechanized vehicles, a mock battle scene, a pushball contest with two 50-man teams, and four tugs of war. Army bands provided a steady stream of music for the show. One amusing competition was the Reveille Contest, in which eight soldiers stripped down to their underwear shorts and sat on cots in the middle of the baseball field. At the sound

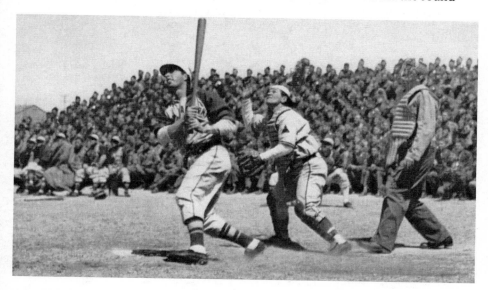

Joe DiMaggio connects with the ball at Camp Cooke. June 1943. Sixth Armored Division booklet *A Pictorial History of the Super 6th.*

of a bugle, they raced to see who could finish dressing first. For participating in this madcap event, the first place winner won a carry bag, followed by the second- and third-place winners receiving a sweatshirt and a tie, respectively. Hollywood movie star Richard Dix skipped his scheduled appearance as Camp Cooke's "dad" for the day.[58]

On July 10, 1943, Larry Crosby, Bring Crosby's older brother and business manager, brought a sensational cast of thirty-nine performers to Camp Cooke. Among the entertainers were Fifi D'Orsay of stage and screen, John Scott Trotter and his orchestra, chanteuse Trudy Erwin, the singing foursome Music Maids, the two singing Haines Sisters, vocalists June "Kit" Carson, Johnny Matteson and his eight girl dancers, conga dancer Zelda, contortionist Margaret Lee, singer of country and western music Johnny Marvin, and "Arkansas Slim" Andrews playing rustic country tunes—all squeezed into two-and-a-half hours of non-stop entertainment.[59]

The musical variety show *Swing's the Thing* was probably the first all-black cast to perform on the Camp Cooke stage. USO–Camp Shows presented the revue in two evening performances at the Sports Arena on July 16–17, 1943. Led by Al Sears and his 13-piece orchestra, with soloist Kenneth Preston, the show also featured the comedy team of Chuck and Chuckles (Charles Green and James Walker), a rhythm-tap dance act by the Three Poms (Ludie Jones, Sylvia Warner, and Geraldine Ball), the comedy of Glenn and Jenkins, and the superb vocals of blues singer Irene Wiley. The shows played to two full houses.[60]

On July 24, 1943, the Sports Arena rocked with entertainment from screen and radio personages. Emceed by William (Billy) Benedict, whose long film career began in the 1930s, the show included two character actors from the *Fibber McGee and Molly* radio show, Arthur Q. Bryan and Bea Benaderet. Bryan performed a comedy skit, and was the original voice of Elmer Fudd. Benaderet performed celebrity voice impersonations. In the 1960s she would have a starring role in the television series *Petticoat Junction* as Shady Rest hotel owner Kate Bradley, and in *The Beverly Hillbillies* as Jed Clampett's cousin Pearl Bodine. The Boyd Sisters (Edith, Elena, and Mildred) sang several south-of-the-border tunes costumed in Central American attire. Years later they would become a popular trio called the Del Rubio Triplets. Actress Barbara Felker strolled through the audience with her accordion serenading servicemen, and later in the show, attired in a tantalizing costume of the Pacific Islands, gave a demonstration of the hula. Sara Berner, a talented dialectician known for her popular radio, cartoon, and movie shorts, had the audience rolling in their seats with her imitations of leading actresses. She wound up her act in a comedy skit with her husband (and manager of the Hollywood

Canteen) Arthur Solomon. The other performers in the show were radio vocalist Shirley Wolcott and two child stars. Billy Roy, who appeared on several radio shows and performed in Hollywood films, played piano and sang to his compositions; and eight-year-old Shirley Collier, dressed in a Scottish motif, danced and sang patriotic tunes. The 69th Armored Regiment band provided the musical background.[61]

Aficionados of classical music sat in rapt attentiveness as Mary Van Kirk, contralto performer at the New York Metropolitan Opera, gave two recitals at Camp Cooke's two service clubs on July 29, 1943.[62]

August 1, 1943, saw the opening of a USO club near the railroad station at Surf, with Faye Porter appointed as its director. Salvation Army Adjutant Cecil L. Day, director of the Walnut Avenue USO club, and Lt. Col. William S. Fowler, executive officer of SCU 1908 at Cooke had conceived the idea for the club. For more than a year they diligently worked to convince a skeptical USO organization that a clubhouse was needed to give entertainment and meals to servicemen who waited for their trains at the Surf depot. The club started as an extension of the USO club on Walnut Avenue in Lompoc, but matured into a separate operation on March 1, 1944. Being one of the few USO facilities in the United States to operate on a military installation during the war, the club's easy access and warm atmosphere made it a popular meeting place for servicemen at Camp Cooke, the B-30 radar station near Arlight, and the Point Arguello Coast Guard lighthouse and rescue stations, and railroad employees. A movie twice a week, frequent dances, and other programs made one forget they were ten miles from town. In its first nine months of operation (August 1, 1943, to April 25, 1944), the club reported an attendance record of 128,000 servicemen.[63]

The club opened in a former mess hall donated by the Army. Engineers from the 5th Armored Division at Cooke, with Coast Guard assistance, turned the building into a welcome retreat. When the Surf club opened, the Lompoc community rallied with donations of furniture, games, recreational equipment, and other articles that brought comfort and joy to service members.[64]

Operating a USO club on Camp Cooke had its unusual moments. During nighttime troop maneuvers the cacophony of tanks and other mechanized vehicles could be heard rumbling past the clubhouse. When artillery opened fire, the building would shake, and dishes occasionally tumbled to the floor.[65]

Closing hour at the club was normally 11:00 p.m. but when troop trains arrived late at the Surf station, sometimes at 4:00 and 5:00 in the morning, the staff stayed on to greet the men with fresh hot cups of coffee and snacks. Sometimes while the men were on maneuvers they slipped in for a quick cup of coffee and a sandwich.[66]

The USO clubhouse at Surf. Faye Porter is standing in the doorway. Courtesy the Lompoc Valley Historical Society.

Dances were regular events at the Surf USO club. Courtesy the Lompoc Valley Historical Society.

With wartime rationing in effect on many essentials, and everything else seemingly unavailable, Faye Porter and her assistants, like so many other Americans, pulled together for the war effort and coped with the deprivations. Most drivers were limited to three gallons of gasoline a week. Porter curtailed the number of trips she made into Lompoc ten miles away and essentially moved into the club with a few other women volunteers. They cooked their own meals and at night slept on cots. Later, when additional buildings were constructed next to the clubhouse, including a combination auditorium/game room, they moved into one of the new dormitories. Tragedy struck on December 9, 1944, when a fire ripped through the dormitory where Faye Porter, Nellie Rita, Velma Rita, and Olive Hayes Reno were sleeping. They narrowly escaped in bare feet and pajamas, losing all their belongings in the blaze.[67]

On August 5, 1943, the Hancock Music Ensemble played another of its classical concert series for Camp Cooke soldiers in Service Club No. 2. Joining the ensemble for its tour of camps for the first time was Belva Kibler, a mezzo-soprano of operatic fame.[68]

Later that month, on August 21–22, the 559th Ordnance Tank Company

Volunteer hostesses at the Surf USO club. Faye Porter is fifth from the left. Courtesy the Lompoc Valley Historical Society.

sponsored a weekend party at Cooke that had as their invited guests thirty-nine lovely young women employees from Lockheed Aircraft Company in Burbank, California. The women traveled to the camp on a company bus with two chaperons. At Cooke they danced at the recreation hall, enjoyed a beach party at Surf, and attended chapel services, all under the watchful eyes of their chaperons. The Army provided all meals and quarters. The wives of the married men in the company were also guests of the Army for the weekend.[69]

On August 25–26 the Camel Caravan musical variety show returned to Camp Cooke for two evening repertoires at the Sports Arena, and an afternoon show at the Red Cross recreation hall for patients and staff at the Station Hospital. Joey Rardin emceed the show that included acrobatic dancer Darlene Ottum, star songstress Kay Carroll, singer Ellen Sutton, Russell and Renee in a trampoline act, and the "Prairie Pioneers"—eight cowboy singers and musicians, with their soloist Bill Hall.[70]

The musical comedy *Passing Parade* closed out the month on August 30–31 with performances at the Sports Arena and Station Hospital. Presented by USO–Camp Shows Inc., the show opened with a chorus line of fourteen girls, followed by the two comedy teams of Marty Collins and Harry Peterson, and Joe Morris and Dorothy Ryan. Other acts included the Lane Brothers acrobatic team; dancer Helene Denizon; Senor Carlos and his harmonica-playing caballeros; the Four Macks, a two man and two women world famous stunt roller-skating team; and singer Li Tei Ming, wife of Charlie Low, owner of the internationally known Forbidden City nightclub in San Francisco.[71]

Actress Jane Wyman, married to Ronald Reagan, was the invited guest of the 280th Field Artillery Battalion on September 2, 1943. Wyman sang at the battalion recreation hall and was made an honorary member of the 76th Armored Medical Battalion.[72]

On September 5, 1943, a film crew from Warner Brothers Studios returned to Camp Cooke to film a movie short that depicted Filipino cultural activities, including games, dances, and favorite cuisine. The movie's theme was built around the traditional lechon meal. More than thirty soldiers from the 2nd Filipino Infantry Regiment and six young Filipino women, led by Pacita Todtod, participated in the movie. Todtod was an activist who at the onset of World War II fought to allow Filipino-Americans to serve in the U.S. Armed Forces. World champion archer Howard Hill directed the film.[73]

The El Paseo nightclub in Santa Barbara brought its festive *Grand International Revue* to the main stage at Camp Cooke on September 6. The song and dance variety show featured a cast of thirty-five entertainers. It came to the camp courtesy of Russell Smith, the cafe manager, in cooperation with the USO club in Santa Barbara and the camp's Special Services Office at Cooke.[74]

In a fun-filled weekend, seventy-four USO hostesses from Los Angeles were the guests of four units of the 6th Armored Division at Camp Cooke on September 11–12. The Division's Special Services Office and the USO arranged for the visit. The women arrived by bus on Saturday evening. After freshening up in their quarters, they were split into four groups and taken to dinner at the four unit mess halls. After dinner, the groups reassembled at the supply battalion's recreation hall for dancing.[75]

On Sunday morning, following religious services and breakfast, the women were taken to the motor park for a flamethrower demonstration and rides in every type of Army vehicle, from jeeps to tanks. The entire group then drove to a firing range where several of the women tried their hand at 30 caliber machine guns. Later that evening it was more dancing and refreshments at the recreation hall before departing that night for Los Angeles.[76]

In separate 70-minute shows, USO–Camp Shows' *WLS National Barn Dance* provided two evening performances at the Sports Arena, and a matinee at the Station Hospital, on October 25–26, 1943. The show had taken its name from the Chicago radio station call letters and its popular radio program of the same name, but had almost no resemblance to its country music theme. In the USO show, Thelma Gardner emceed and sang several romantic songs. Following her were magician Al Nichols, soloist Bonnie Reed, twins June and Julie Watkins singing duets and tap dancing, the Seror Brothers in a comedy pantomime routine, and the novelty act of Bink Mangan attempting to play cowboy music on homemade instruments. A five-piece band from the 382nd Antiaircraft Artillery Battalion at Cooke provided the accompanying music for the show.[77]

On a tour of military camps from San Diego to San Francisco, sponsored by the Hollywood Victory Committee, actor Preston Foster stopped at Camp Cooke for three days on October 27, 1943. Foster greeted hundreds of soldiers and civilians in their offices, shaking hands and signing autographs. In the evenings he met with hundreds more at the service clubs and theaters. For the GIs, Foster came across as a regular guy with no pretenses when he donned a pair of coveralls and helmet, and, carrying a rifle, joined a group of infantrymen in an attack on "Nazi Village," the Army's mock training facility.[78]

Faye Porter, director of the USO Club at Surf, brought another of her many fine shows to the camp on November 7, with an all-girl revue of thirty dancers, singers, and musicians. Included in the chaperoned troupe were the Faye Porter "Personality Girls," and girls from Santa Barbara State College, and the Lompoc and Ventura high schools. The show was presented in two back-to-back 60-minute performances at both service clubs.[79]

On November 8, 1943, world heavyweight boxing champion Joe Louis returned to Camp Cooke for six days. As part of an Army-directed tour of camps around the United States, Louis was leading a quartet of fighters that staged exhibition bouts. The extended stay at Cooke afforded Louis and his fellow fighters time to rest during a grueling schedule that started on August 30.[80]

On November 12, Louis held a three-round exhibition at Camp Cooke with his old sparring partner, George Nicholson. Accompanying them and putting on a short exhibition were boxing champions "Sugar" Ray Robinson and Jackie Wilson. Opening the main show before a wildly enthusiastic crowd of soldiers were four bouts involving pugilists from the 6th Armored Division.[81]

USO–Camp Shows' *Showing Off* played a two-night stand at the Sports Arena and a matinee at the Station Hospital on November 8–9, 1943. According to the *Cooke Clarion*, the best three acts of the show were the comedy of Danny Morton and Sid Ausley, the dance team of Harrison and Carroll, and the singing of the Brian Sisters (Betty, Gwen, and Doris), veteran entertainers of early movies and radio programs.[82]

Al Rosen, Hollywood picture and stage producer, and later actor, brought an all-star cast of entertainers to Camp Cooke on November 14, 1943. The show kept the audience cheering for more than an hour. It featured comedian, singer, and all-around showman Jimmy Durante; three separate vocalists, Marie Belmer, Maxine Conrad, and Carol Parker; and singer/song writer and gifted pianist Jack Owens. He sang a medley of his own works, including "High Neighbor," and songs written by other composers. Character actor and master of ceremonies Donald Kerr opened the show with a dance routine. After the show, the troupe made a 15-minute appearance at Service Club No. 2 where they met with soldiers, and Durante again delighted the audience with short ditties.[83]

The troupe arrived at Cooke earlier in the afternoon and rehearsed with the 382nd Antiaircraft Artillery Battalion orchestra, who rendered the musical portion of the show. Later in the day they

The "old professor," Kay Kyser, Georgia Carroll, and Ish Kabibble, top personalities of the *Kollege of Musical Knowledge* radio show that made its second appearance at Camp Cooke on November 17, 1943. U.S. Army booklet *A Pictorial History of the Super 6th.*

were given a tour of the camp, rode in 6th Armored Division tanks, dined at Service Club No. 1, and shared an evening meal with the 342nd Medical Group.[84]

On November 17, soldiers and civilian employees jammed the Sports Arena for two shows of Kay Kyser's *Kollege of Musical Knowledge* radio show, with Georgia Carroll and Ish Kabibble (Merwyn Bogue). Joining the trio on their second tour of Camp Cooke were announcer Vern Smith, and vocalists Harry Babbitt, Sully Mason, Diane Pendleton, and Julie Conway. The evening show was broadcast nationwide to American troops overseas. The program ended with a rousing rendition of "The Star-Spangled Banner" played by the Kyser band, which brought everyone to their feet singing.[85]

USO–Camp Shows Inc. brought the all-black revue *Let's Go* to Cooke's Sports Arena on November 23, 1943. It featured top New York nightclub acts in two sensational performances. On stage were song and dance man Louis Kelsey, who also doubled as emcee; a vocalist trio, the Three Reeves Sisters; vaudeville comedian Sandy Burns; the song and dance team of Derby and Frenchie Wilson; blues singer Vicki Vigal; and juggling comedian George Rowland.[86]

*Going Some*, the latest USO–Camp Shows revue, played a two-night stand at the Sports Arena, followed by a one-night show at Theater No. 1, on December 6–8. Veteran vaudeville performer and humorist George Dunn, later a movie and television actor, was the show's emcee. Also on stage were the slapstick comedy team of Sylvia and Clemence; the Morrell Trio, a mother, father, and daughter family roller-skating act that was followed by a solo singing performance by the daughter, Beverly; a second singer, Jean Gary; and finally the Two Black Crows, a blackface vaudeville team, with George Moran talking in exaggerated black dialogue.[87]

On December 7, 1943, Twentieth Century–Fox cinemas began filming scenes at Camp Cooke for a movie tentatively titled *I Married a Soldier.* The cast included Frank Latimore, Jeanne Crain, Eugene Pallette, Mary Nash, Stanley Prager, Jane Randolph, Blake Edwards, and, making her film debut, singer Gale Robbins. A supporting cast of about thirty other actors and approximately three-hundred men from the 6th Armored Division participated as extras in the production of the movie. The film was released in October 1944 and titled *In the Meantime, Darling.*[88]

Beryl Wallace took the cast of her radio program *Furlough Fun* to Camp Cooke and presented two shows before a full audience at the Sports Arena on December 17, 1943. The second show was broadcast over radio. The cast members consisted of Spike Jones and his City Slickers band, who had a big hit in 1942 with the song "Der Fuehrer's Face," which ridiculed Hitler; comedian George Riley; the Nilson Twins singing comedy songs; and announcer Larry

Keating. The show's usual slapstick musical comedy turned serious when Wallace interviewed four combat veterans from Guadalcanal. Gilmore Oil Company sponsored the program.[89]

The USO–Camp Shows production of *Come What May* appeared at Cooke on December 20–22, 1943. The first two nights it played the Sports Arena and then moved to Theater No. 1 for the third night. The cast members included vocalist Carol Dexter; puppeteer Catherine Westfield and her near life-sized puppets modeled after celebrities of the day; the Three Debs (Lila, Barbara, and Peggy) performing modern ballet; the vaudeville team of Lynn Russell, who also emceed the show, and Marion (or Miriam) Farrar performing comedy skits; and hypnotist Howard Klein. Klein brought twenty-five soldiers on stage and hypnotized them to do everything from shooting dice to minding a baby. Pianist Henry Fluegge with the show, and the 6th Armored Division band, supplied the music.[90]

On December 25–26, 1943, the Hollywood Victory Committee brought plenty of holiday cheer to service members at Camp Cooke with a comedy show filled with song, dance, and pratfalls. Its featured players were actor and comedian Edgar Kennedy, master of the slow burn comedy routine, and comedian Harry Martin with his ukulele. The other performers included vocalists Delores Peterson, dancer Barbara Gay, singer Romenta Barnett, and entertainer Jim Burke. All made encore performances to cheering crowds.[91]

Early Saturday evening, New Year's Day 1944, two busloads of young women from the Desert Battalion descended on Camp Cooke on an invitation from the 31st Medical Group. The ladies attended a dance later that evening at the group's Field House on Wyoming and F Streets. The battalion's entertainment squad dressed in western garb and, backed up by an Army band, put on a show that included comedy routines, romantic songs, parodies, and accordion specialties such as "Pistol Packin' Moma," played by Beverly Warren. Special guests appearing on the entertainment program were Freddy Mercer, a 12-year-old singer who played the part of LeRoy in *The Great Greensleeve* radio program; June Carlson, former feature player in the Jones Family comedy movies, who did a monologue; and Beverly Brown, who appeared regularly on the Edward G. Robinson weekly radio show. The evening concluded with a community sing-along of sentimental favorites. The next morning the ladies attended religious services then piled into jeeps and ambulances for tours around the camp. Following an afternoon dance and final goodbyes, the ladies headed home to Los Angeles and later to the next camp.[92]

The Desert Battalion returned to Cooke three months later, on March 11, 1944, to help the 28th Antiaircraft Group celebrate the first anniversary of its activation. Forty-five battalion girls enjoyed a Saturday night dinner,

Individuals of the Desert Battalion who appeared on the program at the dance of the 31st Medical Group in their Field House New Years evening are shown in the upper picture. From left to right: Lucille Hart; June Carlson, former feature player in the Jones Family comedy movies who did a monologue; Beverly Warren, who presented several accordion numbers; Freddy Mercer, boy soprano traveling with the desert girls, who played the part of LeRoy in *The Great Greensleeve* radio program; Francine Horbach; and Gladys Robinson, the "brigadier" of the "Battalionettes." Lower picture shows the start of the dance when the ladies were lined up on one side of the hall and the men on the other; at the "go" signal each side found their dancing partner. U.S. Army/*Cooke Clarion*.

dancing, and a floor show with the boys. The following morning it was chapel services, jeep rides, and a movie party at Theater No. 3. At a final anniversary banquet later that day, group commander Col. Allison Scott gave the ladies the honor of presenting military Good Conduct ribbons to the men who earned them; and, of course, each ribbon came with a good luck kiss that brought lots of cheers from the crowd. Attending much of the activities were special guests Mr. and Mrs. Jack Hope, brother of entertainer Bob Hope, and actor Ned Sparks and his wife.[93]

The Desert Battalion was established in the late spring of 1942 by Gladys "Robbie" Robinson, the wife of actor Edward G. Robinson, to bring female companionship to homesick soldiers training at remote desert camps in California. At these camps General George S. Patton prepared his troops for the harsh conditions they would face in the deserts of North Africa.[94]

As the Army requested more visits from the Desert Battalion, additional training camps in California and in the Arizona desert were added to the battalion's itinerary. More than six-hundred women, most unmarried and all between the ages of 18 and 25, participated in the Desert Battalion. On weekdays they might work in offices, factories, and war plants, but on weekends they were chaperoned on bus trips to these remote camps. In its two years of existence, the Desert Battalion covered 350,000 miles and entertained about 250,000 men.[95]

On January 14–15, 1944, USO–Camp Shows Inc. presented *What's Next* in two evening repertoires at the Camp Cooke Sports Arena. The featured players were comedian-juggler Trout Taylor and his wife Mickie, magician Paul McWilliams, hula dancer Paula Peiry, the vocals of the "Three Tones," tap dancer Johnny Curtis, and Jack Waldron as emcee.[96]

During its second year of hosting entertainment shows at military installations around the country, the Shell Oil Company for the first time brought its Shell Military Show to Camp Cooke on January 25–26, 1944. Two 90-minute evening performances were held at the Sports Arena, with a matinee at the Station Hospital. The cast of six players were magician Mariana; vocalist Bonnie Brier; dancer and acrobat Ida Lynn; musicians Lloyd Simpson on piano and Johnny O'Brien playing harmonica; and ventriloquist Lucille Elmore with her two dummies, Andy and Snoopie.[97]

Crooner Rudy Vallee, already a popular star of radio, screen, and stage, and now a lieutenant leading the 11th Naval District United States Coast Guard band, presented an afternoon concert at the Sports Arena on February 1, 1944, to more than 5,000 soldiers and civilian employees. Vallee was on a tour of military installations to help promote the sale of war bonds. The 90-minute program featured a mixture of vocals and instrumentals. Robert Maxwell, eminent composer and harpist, and a member of the Coast Guard band, played several encore performances. Singing several popular romantic ballads was Jane Frazee, actress, singer, and dancer. A jovial touch was added by the "Coast Guard Cutters," a three-man tumbling act from the vaudeville stage, and storyteller Dave Wade.[98]

**Rudy Vallee staged an afternoon concert at Cooke to promote the sale of War Bonds. February 1, 1944. Courtesy the 11th Armored Division/Bob Peiffer.**

On February 7–8, USO–Camp Shows Inc. presented the variety show *Say When* at the Cooke Sports Arena and Station Hospital, which had audiences applauding and cheering. Tapping away on the xylophone were the Musical Johnstons, followed by the wonderful dance moves of Eve Matthews, the harmonizing voices of the Three Dixon Sisters, and, in three separate comedy acts, Charlie Bohn, Henza Venton, and the comic acrobats Emmet Oldfield and Frank Rooney. Musicians from various field artillery bands at Cooke provided the obbligato accompaniment for the show.[99]

An all-star cast of Hollywood entertainers performed at the Sports Arena in the Collinstone Revue, which featured actors Charles Collins and his wife Dorothy Stone. The February 13 show included singer Dorothy Cordray; another songstress, Heidi Olson; dancer Martha Scott; actresses June Knight and Mary Brian, with the Rhythm Rascals band; and musical barbershop harmonies by the quartet of Pryor Bowen, Marek Windheim, Milton Parsons, and Neil Hamilton, accompanied by pianist Tommy Griselle. A generation later, TV viewers would come to know Hamilton as Police Commissioner Gordon in the campy 1960s *Batman* series. Lorena Wade Fletcher of Los Ange-

A lucky sergeant is the center of attention for Dorothy Stone (second from left) and three showgirls. February 13, 1944. Courtesy the 11th Armored Division/Bob Peiffer.

les, who organized talent groups to visit camps in the L.A. area, worked with the camp Special Services Office to bring this show to Camp Cooke.[100]

General Electric's "House of Magic" show returned to Camp Cooke on February 23–24, 1944, exactly one year after its initial presentation. The "magic" consisted of glimpses into futuristic technologies that included showing a man walking away from his own shadow and a miniature electric train obeying voice commands.[101]

Renowned pianist, arranger, and composer Ferde Grofé, who began his career with the Paul Whiteman Orchestra in 1920 and later composed his most famous work, "Grand Canyon Suite," was the star attraction at Camp Cooke on February 27, 1944. Teaming up with local musical talent at Cooke, the girls' quartette of Santa Maria Union High School, and stage, screen, and

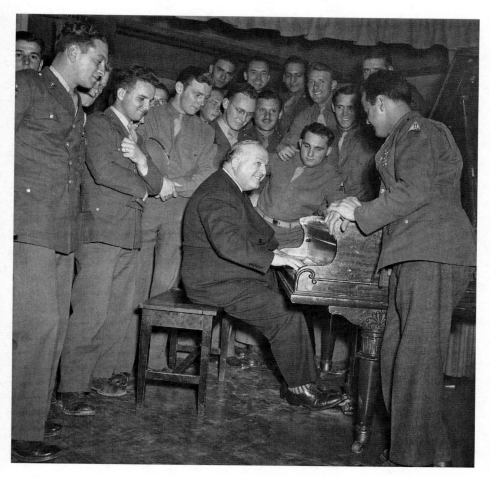

Renowned pianist, arranger, and composer Ferde Grofé giving an impromptu concert for a crowded gallery after performing two shows at Service Clubs 1 and 2. February 27, 1944. Courtesy the 11th Armored Division/Bob Peiffer.

radio singers, Grofé packed both service clubs with appreciative audiences. The same show had been presented at the high school the previous day to help promote the sale of war bonds. Cadets from Hancock College of Aeronautics in Santa Maria sponsored the shows.[102]

Playing to a near capacity audience in the Sports Arena on March 8–9, 1944, and a matinee at the Station Hospital, was *Tell Me More*, a variety show presented by USO–Camp Shows Inc. The emcee for the show and one of its top performers was Viola Layne. An impressionist singer from the musical comedy field, Layne had a remarkable range in her singing voice and all the mannerisms of the stars she imitated. Layne became Bonnie Baker, Virginia O'Brien, Shirley Temple, Carmen Miranda, and Gracie Fields all in one night. Rounding out the ticket were the sultry songstress and dancer Gloria Manners, xylophonists Eddie Blum, the crazy antics of comedians Les and Poppy Lamarr, and finally the blackface vaudeville team of the Two Black Crows. The show had opened and closed with the six shapely and internationally famous Brucettes performing syncopated acrobatics.[103]

On March 26, 1944, soldiers again jammed the Sports Arena, this time to attend Lorena Wade Fletcher's *Stars of Tomorrow* show, with special guests actor Dana Andrews and vocalist Gloria Delson. Showcasing predominantly female singers and dancers on stage, the acts drew vociferous rounds of applause and cheers from the almost all-male audience.[104]

The following month, on April 5–6, USO–Camp Shows brought to the Sports Arena *Funny Side Up*. This show featured Bud and Rosa Carlell, a husband and wife team doing rope and whip tricks; Carl Sheldon and his two sisters in a tumbling and acrobatics act called "Gob and Two Gals"; Evelyn Kenyon on xylophone; dancer Charlotte King; and comedians Ben Drohan and Mary Dupree.[105]

Rudy Vallee and the 11th Naval District United States Coast Guard band returned to Camp Cooke on April 24 to give another solid musical entertainment show. The musical selection varied from contemporary Big Band Jive to George Gershwin, and a sprinkling of nautical tunes. Band member Robert Maxwell played a number of solos on the harp. Plenty of comic relief was added to the show in the form of a tumbling team called the "Coast Guard Cutters" and storyteller Dave Wade.[106]

USO–Camp Shows' *Around the Corner* played before two full audiences at the Sports Arena on May 4–5, 1944. The show opened with a medley of music from the 11th Armored Division band. The individual acts included Hap Hazard (Harold Hart) and his wife Mary performing amusing juggling tricks, Eddie Manson playing harmonica, slapstick comedy from York and Tracy, dancer Barbara Champeon, and the vocals and piano playing of Chenault and Day.[107]

Former world heavyweight boxing champion Jim Jeffries, and former world welterweight champion Jimmy McLarnin, guest refereed at a pugilist tournament held at the Camp Cooke Sports Arena on June 7, 1944. Joining an audience of soldiers watching the action were more than fifty Lompoc high school students. The students were invited guests courtesy of the 11th Armored Division and the Lompoc Sports Association, which arranged for their transportation and admission.[108]

On June 8–9, 1944, USO–Camp Shows Inc. presented *All Is Well* at the camp Sports Arena. The show featured comedians Collins and Peterson, who also doubled as emcees; Marie Belmer singing several numbers and executing a few dance steps that elicited plenty of whistles from the spectators; Emmet Oldfield and Frank Rooney, comic acrobatics; the Johnny Pastime Trio, a man and two women performing ballroom dancing; Whitey Field, comic juggler; and Edna Mae Kenyon on xylophone. The 11th Armored Division band provided the music for the show.[109]

A few days later, on June 18, the *Clambake Follies* opened at Camp Cooke. The second of two variety shows sponsored by Bing Crosby, it was brought to the camp by his brother Larry Crosby, who emceed the show. The revue opened with the Johnny Bushallow girl dancers in sarongs performing a hula number. Next came cowboy singer Shorty Mavin, the comedy juggling team of Johnson and Diehl, harmony singing by the Brian Sisters, precision roller skating by the Morrell Trio, singer June "Kit" Carson, another vocalist named Delores, and finally a trained dog act. The show closed with a chorus performing a patriotic number dedicated to Camp Cooke. The 96th Infantry Division band from Camp San Luis Obispo supplied the music.[110]

Fortunio Bonanova, the Spanish-born singer and actor who appeared in

Fortunio Bonanova, backed up by the 11th Armored Division band, the Cobandos, sings one of his hit songs. June 1944. Courtesy the 11th Armored Division/Bob Peiffer.

several Hollywood movies, including *Citizen Kane* and *For Whom the Bell Tolls*, was the special guest of the 11th Armored Division at Cooke on June 24–25, 1944. Bonanova performed at Service Club No. 1, and later at the Division's Officers' Club, singing many of the songs he made famous in his screen performances. The following day he entertained patients at the Station Hospital.[III]

Pfc. Andre Arceneaux, 21st Armored Infantry Battalion, watches anxiously as Orson Wells, impresario of the radio show which played at Camp Cooke on July 12, 1944, demonstrates magic as he prepares to "amputate" the soldier's hand. Courtesy the 11th Armored Division/Bob Peiffer.

**Susan Hayward signing autographs. July 12, 1944. Courtesy the 11th Armored Division/Bob Peiffer.**

*Brazilian Nights*, a south-of-the-border stage revue featuring six acts of singing, dancing, and comedy, opened at the Sports Arena on July 10–11. This USO–Camp Shows production included Sylvia and Margo performing international dances, vocalist Gracie Scott singing songs from South America, the comedy skits of Rolando, pianist Mel Morris, the rough and tumble comedy of McFarland and Brown, and Al and Joan Allen incorporating acrobatics and pantomime into their dance act. Members of the 259th Army Ground Forces band provided the music for the revue and played several numbers before the show opened.[112]

On July 12, Orson Wells presented his *Mercury Wonder Show* from the stage of the Sports Arena. Among the show's highlights were a dramatic reading by Wells from William Shakespeare's *King Richard II*, a comedy skit featuring Wells and actress Susan Hayward, and magic tricks performed by Wells. An audience of more than 4,000 people gave a rousing reception to the show, which was broadcast nationally.[113]

**Hollywood starlets and filmland's ace entertainers gave their time and talent to enter-tain soldiers at Camp Cooke on August 15, 1944. Left to right: Peggy Stewart, Bonita Granville, Phyliss Ruth, and Shelia Ryan. Courtesy the 11th Armored Division/Bob Peiffer.**

On August 9–10, 1944, USO–Camp Shows Inc. presented at the Sports Arena a Harlem stage revue called *What's Cookin'.* The show featured music by a 10-piece orchestra, with Alice Tyson leading the vocals and a chorus line of six dancers. Also headlining were vaudeville-style comedians Bert Howell and Buddy Bowser, jazz and blues vocalist Laurel Watson, jazz musician and singer Herbert (Herbie) Cowan, tap dancer Danny Alexander, and ventrilo-quist Jester Calhoun. Although well received by the audience, the camp news-paper reported the show in derogatory and racist terms.[114]

For three hours, on August 15, another USO–Camp Show rocked the Sports Arena by generating laughter and applause. Appearing on stage were comic actor Edgar Kennedy and his wife; actors Ray Bolger, Henry O'Neill, Fred Sanborn, Bonita Granville, Phyliss Ruth, Shelia Ryan, Wallace Ford, Peggy Stewart, and John Carroll, currently a captain in the Army Air Force; prize fighter Jimmy McLarnin; comedian Charlie Kemper; vaudeville come-

**Left to right: Fred Sanborn, Wallace Ford, Jimmie Fidler, and Ray Bolger. August 15, 1944. Courtesy the 11th Armored Division/Bob Peiffer.**

dians Rose King and her husband Chick Yorke; and three attractive women dressed in black evening gowns playing harps and calling themselves the "Three Harps of Swing." Gossip columnist and radio-personality Jimmie Fidler emceed the show. Earlier that day, and the following day, many of the entertainers visited with hospital ward patients.[115]

Crowds gathered outside the Sports Arena on September 7, 1944, waiting to get inside to hear Jimmy Dorsey and his band, with vocalists Teddy Walters and Anita Boyer. The show played over a national radio network. Arrangements for the evening's performance were made by the 11th Armored Division's Special Services Office, and was sponsored by the Coca-Cola Bottling Company's Victory Parade of Spotlight Bands.[116]

In two evening performances at the Sports Arena on September 8–9, USO–Camp Shows Inc. presented *It's a Pleasure*. The troupe include comedian and harmonica player Al Mardo, vocalist Andrina, lady magician Flauretta, country comic Charles Withers, ballroom dancers Montez and Maria, and feats of strength demonstrations by Dave Winnie. Using a portable trapeze

**Comedian Edgar Kennedy clowning with 11th Armored Division MPs. Left to right: Cpl. John McGee, Pvt. James Poarch, Cpl. Maurice Zelter, Pfc Edwin Pasneiski (shaking hands), and Sgt. Paul Monahan. Courtesy the 11th Armored Division/Bob Peiffer.**

and bars, Winnie hung by his knees from a crossbar fifteen feet above the stage while supporting his assistant, Dolly McManus, with his teeth as she spun on a ship's anchor.[117]

"Loaded with laughs, thrills and plenty of girls" is how the *Cooke Clarion* described the October 9 show *Hold Tight*, which filled every seat in the Sports Arena. Best known among the USO–Camp Shows' troupe of fifteen players was Sam Hearn, who played the "Schlepperman" character on Jack Benny's radio program. The bill's other performers were the novelty dance routines of the six Catherine Behney Girls, who opened with a French *can-can*; the harmonies of the Keller Sisters (Annie Catherine "Nan" and Kathryne Ann "Taddy"); Joe Lane and Pearl Harper singing humorous parody songs; the Nathan Brothers, combining their acrobatic and strong man act with the playing of a violin and mandolin; and female acrobatic dancer and dead-pan comic Gerry Wright. The 97th Infantry Division dance band furnished the music for the revue, with Josephine Bond from the show serving as pianist and musical conductor. A matinee was also performed at the Station Hospital.[118]

The Shell Military Show, sponsored by the Shell Oil Company, returned to the Sports Arena at Camp Cooke on Halloween 1944 with six variety acts packed into a one-hour show that delighted GIs and their guests. Its star performer and emcee for the show was ventriloquist Lucille Elmore. Also in the revue were harmonica king Johnny O'Brien, pianist Lloyd Simpson, dancer and acrobat Ida Lynn (who drew loud applause from the audience with her hand-walking and flip-flops), singer Wanda Vaughn, and Carlisle the magician. Earlier in the day the cast gave an afternoon performance at the Station Hospital.[119]

USO–Camp Shows Inc. presented *Thanks Loads* at the Sports Arena on November 8–9. The cast included song and dance comedians Benny Ross and Maxine Stone, who also acted as emcees; the vocals of the "Jive 'n Jills" (Marianne Craig, Joyce James, and Bonnie Nottingham); the acrobatic and tap trio the Wyse Girls (Viola DeNaro, with twins Jane and June Hughes); the graceful ballet of danseuse Roberta Ramon; and magician Mysterious Damon. Music for the revue was provided by the 97th Infantry Division band.[120]

Beginning in late November, Camp Cooke became the backdrop for Columbia Pictures' production of *Counter-Attack*, starring Paul Muni and Marguerite Chapman. The story takes place in German-occupied Russia in 1942. In the scene filmed at Cooke, the Russian army is outside Stalingrad crossing the Volga River in a counter-offensive. Portraying this part of the action were one-hundred soldiers from the 23rd Replacement Depot unit and three platoons of the 782nd Tank Battalion dressed in Russian army uniforms.[121]

With guns firing blanks, American tanks loaded down with infantrymen crossed the Santa Ynez River. To heighten the action during the crossing, inert explosives planted in the river were set off, sending columns of water skyward to simulate enemy artillery fire. To ensure that none of the tanks would become stuck, a shallow section of the river was selected and lined with steel matting.[122]

The river scene was filmed on November 27, 1944, but was reshot a week later on December 4. During the next two days, shorter sequences were also filmed showing the tanks assembling and approaching the water. The movie was released in April 1945.[123]

*Step Lively*, another of the monthly USO–Camp Shows featured at Camp Cooke, played the Sports Arena on December 2. Headlining the one-hour show was Stan Kavanagh, comic juggler and emcee; the five Abdulla Girls in a tumbling and acrobatic number; vocalist Dorothy Crowley; the two Drake Sisters in a tap dancing routine; ventriloquist Sully Ward; and the comedy team of Roger Williams and Alice. Williams imitated everything from Scotch bagpipes to a Model T Ford starting on a cold day, while Alice played the accordion.[124]

When the troupe arrived two hours after the scheduled opening, the staffs of the two service clubs kept the audience in place by pulling together an impromptu GI talent show of singers, musicians, and comedy acts that was enthusiastically received. An international note was added to the show by Pvt. Salvatore Esposito, 142nd Italian Quartermaster Service Company. Technically a prisoner of war, he delighted the audience with his singing of popular Italian tunes.[125]

Opening a one-week engagement at the Sports Arena on December 10, 1944, was Benny Fox's "Star-Spangled Circus." Headlining the show was La Tosca. Her real name was Tosca Canestrelli, and she was a member of that famous Italian circus family. La Tosca was known as the Queen of the Circus for her regal beauty and spectacular act on the bounding rope. She would kneel, sit, and lie prone on a rope strung between two steel poles. From a sitting position she would perform back flips and other amazing maneuvers. She was the world's only female bounding rope performer at that time, and today is still the only woman in history to have accomplished the double backward somersault on the bounding rope. The other acts included the Charles Siegrist troupe of flying trapeze artists; top clown Ernie Wiswell; Eric Philmore, the world's premier juggler; Mickey King, aerial sensation; August Jansley performing acrobatic antics atop a 210-foot pole; Maxima King of the slack wire; the Landos little people; and a myriad of animal performers. Music for the shows was provided by a 30-piece band from the 97th Infantry Division at Cooke.[126]

USO–Camp Show Inc. kicked off New Year's 1945 at Cooke on January 12 with a musical revue called *Hats Off*. The seven-act show featured juggler James Evans, who kept various ill-assorted items (including a 95-pound bed) perfectly balanced in mid-air; the "Park Avenue Sextets," a precision dance team that performed a tap-chorus routine followed by a medley of South American movements, and ended with a rope-skipping routine; the comedy piano and singing vaudeville duo of Harry LeVan and Lettie Bolles; the vocals and music of Bill Herron on xylophone and Bill Richardson on guitar; Al and Joan Allen, with celebrity impersonations ranging from famous dancers to Abbott and Costello routines; and concert soprano Olga Dubinetz singing popular tunes. The 97th Infantry Division band opened and closed the show with "jump" tunes, and provided the musical accompaniment for individual acts.[127]

On February 12, 1945, USO–Camp Shows Inc. presented the musical comedy revue *Who Goes There* at the Sports Arena, and a matinee performance at the Station Hospital, which drew loud applause. The acts featured comedian and singer Jack Leonard, who also doubled as the master of ceremonies; the Garri Girls, a six-girl chorus line; musicians Ferrari and De Costa playing a medley of tunes; vocalist Mary Lee Carroll; the comedy cartwheels, flips, and tumbling of the Emerald Sisters; singer and dancer Annette Ames; and Beverly Peterson as accompanying pianist for the show.[128]

Another USO–Camp Show, called *Perk Up*, was performed on April 12 and included the comedy of Benny Ross and Maxine Stone; the synchronized tumbling of the Hoffman Sisters; a harmony duo in the form of the Cleveland Sisters; and the graceful ballet of Helene Denizon. Rounding out the one-hour show was radio's effervescent Kay Parsons. She sang two solo numbers and several other songs accompanied by members of the Italian Service Units at the camp.[129]

On May 10, 1945, three days after the surrender of Germany that brought the war in Europe to a close, USO–Camp Shows presented *Broadway Maneuvers* at the Sports Arena. Among the cast of dancers and comics were the velvet-toned Keller Sisters in a harmony duo; sisters Flo and Jo Cook playing popular and semi-classical music on the xylophone; Monti and Lyons on guitar and mandolin; the singing and dancing of the Gray Sisters; and Bob and Helene Ranous performing acrobatics.[130]

*Hello Joe* was the next USO–Camp Show to open at the Sports Arena, on June 9. On the ticket were the five Elaine Seidler girls in a high-stepping chorus line; the acrobatic team of Dean and Brown; the singing Fisher Sisters (Jean and Jeffie); and the comedy of Smith, Rogers, and Eddy—two men dressed in disheveled clothing and derby hats playing to an attractive woman (June Rogers) in the role of girl foil (this comedy pantomime dance act reportedly kept the audience in hysterics). Milton Douglas and a woman named Priscilla shared the master of ceremonies role and received plenty of laughs with their playful comic routines. On piano for each of the acts was George Hackett. Since most soldiers in the camp were busy preparing for a headquarters inspection the following morning, the audience of several hundred was largely composed of civilians and Italian Service Unit personnel.[131]

USO–Camp Shows Inc. brought *Chicks and Chuckles* to the Sports Arena on July 10. The variety show opened with a six-girl chorus line. The featured singer, Muriel Lane, was previously with the Woody Herman Orchestra, and later with Bing Crosby on the *Kraft Music Hall* radio program. The other players included the Two Black Crows, with George Moran and Rade Sadler; comic Lee Simmons as master of ceremonies; and a modern dance-ballet team.[132]

The following month, USO Camp Show's *Gee-Eye Revue* played at the Sports Arena, featuring drum player Charlie Masters; vocalist Beverly Knox; the comedy of Ray Parsons and Bob Taylor; Miller and Jean, an acrobatic dance team; and Johnny Hyman, billed as the blackboard wizard, with his quick-firing calculations that he chalked on slate.[133]

On September 10, 1945, a month after World War II had ended, USO–Camp Shows Inc. presented *Have a Look* at the Camp Cooke Sports Arena. The audience witnessed the Great Lester saw his attractive assistant Diane Rivers in half, and by legerdemain put her together to assist in other tricks. Actor and writer Bob White did singing imitations of Frank Sinatra and other contemporary artists; Sylvia and Clemence bounced around the stage in their madcap acrobatic act; vocalist Martha Tarlton sang to the boys in uniform; and Harrison, of the Harrison and Carroll dance team, performed some fancy footwork. Most, if not all, the entertainers in this troupe, including the 64th Army Ground Forces band, also participated in the next day's show, the *Clambake Follies*.[134]

The *Clambake Follies*, assembled by Larry Crosby, the brother of singer Bing Crosby, was a rollicking two and a half hours of audience cheering and applause. Presented at the Sports Arena on September 11, it featured a chorus line of dancers; the Morrell roller skaters; rumba and conger dancer Zedra; two separate jugglers, Ray Wilbert and Serge Flash; the beauteous Luther Twins in a dance act; the musical Bonneys, a husband and wife and daughter team; ventriloquist Sheila Harrington; the singing Brian Sisters; comedian and guitarist Frank Saputo; singer and choreographer John Bushallow (later Jon Gregory); magician the Great Lester and his assistant Diane Rivers; pianist Louis Cheney; and actor and writer Bob White as the show's emcee. The music was supplemented by the 64th Army Ground Forces band.[135]

Bette Davis, Eddie Bracken, Dick Erdman, Johnny Coy, Robert Alda, and Don McGuire were some of the entertainers who participated in a hilarious mix of comedy skits, songs, and dance at the Sports Arena on September 15, 1945. In one skit, Davis graciously took two pies to the face from dancer Johnny Coy. Also performing for the audience of service members and civilians were tap dancer and baton twirler Carol Jean Johnson, songstresses Inez Gorman and Gloria Elwood, magician Roy Benson with sleight-of-hand tricks, and pianist Rosa Linda, formerly with the Paul Whiteman Orchestra. The Hollywood Victory Committee presented the show in two evening performances. The 13th Armored Division band provided the music.[136]

On September 24, 1945, Louis Armstrong, wielding his trumpet and mellifluously leading his orchestra in such tunes as "I Can't Give You Anything but Love, Baby," and "Keep Jumping," packed the Sports Arena with 5,000 GIs and civilians. Joining Armstrong in the music fest was chanteuse Velma Middleton and at least seven other band members.[137]

The Hollywood Victory Committee presented another of its variety shows at the Sports Arena on October 2, this one featuring singer and tap dancer Darlene Garner, dancer Gerri Gale, film comedian Frank McHugh, vocalist Kim Kimberly, and comedian Joe DeRita, who would later join the Three Stooges as Curly Joe. The show included other singers and dancers as well. The 20th Armored Division band provided the accompanying music.[138]

Making his third appearance at Camp Cooke, on October 9, 1945, Bob Hope, along with Frances Langford, Trudy Ewan, Jerry Colonna, Skinnay Ennis and his band, and special guest boxer Billy Conn, played two shows before ebullient audiences. The first performance was at the ballpark across from the Sports Arena on Washington Avenue. The second show was a nationwide broadcast of his radio program from Theater No. 2. Some 10,000 soldiers, WACs, nurses, and patients from the Station Hospital greeted the troupe.[139]

On the evening of October 9, USO–Camp Shows presented *Jolly Times* at the Sports Area. The main acts were Elaine Seidler's Park Avenue Debutantes, a chorus line of five girls; the Le Shonnes, a ballroom dance couple;

The cast of the Hollywood Victory Committee's show on October 15, 1945. Upper left, Arthur Treacher and Bunny Cutler share the spotlight in a comedy skit; upper left center, entire cast leads members of the audience in singing "God Bless America"; upper right center, Robert Armstrong, who acted as master of ceremonies; upper right, Anne Jeffreys, movie actress, who sang "I'll Buy That Dream"; lower left, Gerri Chavey, diminutive acrobatic dancer, prepares to skip rope with one foot behind her head; lower center, Joe E. Brown, featured performer of the show, clowns with members of the audience; lower right, Julie Lynn, singer of Latin songs, gives out with "Rum 'n Coca Cola." U.S. Army/*Cooke Clarion.*

torch singer Ruth Whitney; Johnny Reading as master of ceremonies; and various comedy and singing acts from vaudeville circuits and nightclubs. The 20th Armored Division band covered the music portion.[140]

Six days later, on October 15, the Hollywood Victory Committee delivered a lively musical-comedy variety show that packed the Sports Arena in double two-hour performances. On the ticket were comedian and entertainer Joe E. Brown; actors Arthur Treacher, Robert Armstrong, Anne Jeffreys, and Bunny Cutler; vocalist Julie Lynn; old-time vaudevillian Davie Jamison; Vivian Faye and a cast of dancers; and the O'Brien comedy trio. Just after the trio had completed their act, 53-year-old Thomas Brennan, known on stage as Tom O'Brien, suffered a fatal heart attack off-stage.[141]

The Shell Military Show, sponsored by the Shell Oil Company, returned to the Sports Arena on October 22, 1945, with harmonica player Johnny O'Brien, ventriloquist Lucille Elmore with her two dummies Andy and Snoppie, vocalist Wanda Warren, dancer Ardis May, Carlisle the magician, and pianist Eunice Steel.[142]

The Hollywood Victory Committee brought another of its all-star lineup shows to the Sports Arena on November 8. Cast members included Vivian Faye and Joy Stewart in separate dance acts; actress and vocalist Patricia Mirage; several comedy skits with actors Jack Carson, Janice Paige, Arlyn Roberts, Irene Ryan, and comedian Tommy Wells; and finally Jack Cavanaugh, a versatile performer doing rope spinning tricks. Ryan would receive notoriety for her portrayal of Granny in the 1960s television show *The Beverly Hillbillies*. Music for the show was supplied by the 20th Armored Division band, directed by Ronald Buck, who came with the troupe.[143]

On November 15, USO–Camp Shows Inc. featured actor Sidney Marion and seven other entertainers at the Sports Arena. In the opening number, Julie Ballew joined Marion in a hilarious comedy routine, and sang several humorous ditties; Judy Todd sang and danced; Stan Gilbert and Sid Ausley punctuated their comic patter with a dance routine; The Great Wylito performed magic; Lucille Angel sang contemporary romantic songs; and pianist Joyce Wellington played boogie and background music for other acts, and played with the accompanying 20th Armored Division band.[144]

*You Said It*, a fast-moving variety show presented by USO–Camp Shows Inc., entertained a near-capacity crowd at the Sports Arena on November 23. On stage were Maurice and Tessie Sherman performing a mix of music, comedy, and dance; the Cover Girls, an octet of girl dancers; the Latlip Sisters (Rosaline, Madeline, and Ginger), a dance team combining some rough and tumble antics with the graceful movements of ballet; the duo of Andrew McFarland and Dorothy Brown in a comedy acrobatics act; and vocalist Sally Winthrop.[145]

A week later, on November 29, 1945, another USO–Camp Show production featured actors Chill Wills, Audrey Totter, Marvelle Andre, Billy Curtis, and Jerry Maren in various skits; Davie Jamieson, dance impersonator of old-time vaudeville performers; Kenny Pierce, celebrity impersonator; vocalist Julie Lynn; singer Lynn Lyons; and a pair of vocalists—the Wilde Twins (Lee and Lyn), formerly of the Bob Crosby orchestra and who, in 1943, appeared with the band in *Presenting Lily Mars*, starring Judy Garland and Van Heflin.[146]

USO–Camp Shows' presentation of *Step on It* played before an audience of 1,500 at the Sports Arena on December 21. The sixty-minute variety show featured musical numbers interspersed with humorous patter and dance routines by the Kitty Wolfe chorus line of four pliant girls; the dancing and singing act of the McCune Sisters; the husband and wife comedy team of Fred and Bobbie Brown; and the comedy and dance team of Dick and Evelyn Barclay.[147]

F. Hugh Herbert's stage play *Kiss and Tell*, a family-life comedy that first played on Broadway in 1943 and two years later was made into a movie, was performed at Cooke on January 31, 1946. Sponsored by USO–Camp Shows, the cast of fifteen veteran Broadway actors included Rose-Ellen Cameron and Judson Pratt.[148]

In his second visit to Camp Cooke, on February 12, 1946, Chill Wills appeared alongside a cast of talented performers, including actress Marvelle Andre; country singer and actress Betsy Gay, whose movie roles included the character Effie in the *Our Gang* series; singer Virginia Johnson; contortionist Virginia Carroll; Cliff Arvin and his dancing marionettes; dancer June Rumsey; and musician Jiuonti. Gowned in a slinky silvery evening dress, Jiuonti tossed her long auburn tresses repeatedly while playing "St. Louis Blues" on a trombone with the 20th Armored Division band.[149]

On March 8, 1946, USO–Camp Shows Inc. presented its last show at Camp Cooke, *Town Topics*, to an audience of four-hundred enthusiastic service members and civilian employees. Taking the spotlight was Virgil Whyte's "Musical Sweethearts," a 12-piece all-girl orchestra that began the program with a new version of "Kansas City Blues" under the direction of Alice White. White also doubled as vocalist and drummer. Also on stage were tap dancer Pearl Kay, the acrobatics and tap dancing team of Rosalie and Martha Woodson, and the comic acrobatic team of "Paul Lavarre and brother."[150]

The Shell Military Show returned to Camp Cooke for a final performance at Theater No. 3 on March 29, 1946. The entertainers were harmonica player Johnny O'Brien, vocalist Wanda Warren, dancer Ardis May, Carlisle the magician, pianist Eunice Steel, and master of ceremonies Joe Twerp. This show may have featured the last group of entertainers to perform at Camp Cooke before it closed on June 1, 1946.[151]

Likewise, the future of USO clubs in Lompoc and Santa Maria hinged on the Army's plans for Camp Cooke. When the 97th Infantry Division departed in March 1945 without an equally large replacement unit scheduled for the camp, the USO organization decided in June 1945 to close the clubhouse on South H Street. The Army's announcement in February 1946 that it was closing Cooke prompted the USO to begin issuing termination notices to its remaining clubhouses in the area. Meanwhile, the building it vacated on South H Street remained for the next fourteen years, hosting fraternity organizations until succumbing to the wrecking ball in December 1959.[152]

In Santa Maria the USO club and the dormitory ceased operations about the same time, on March 31, 1946. The city's recreation department took over the clubhouse and turned it into a youth recreation center. The USO club at Surf marked its final curtain call on April 29, 1946, with an evening dance for five-hundred servicemen and civilians. Its four temporary buildings stood vacant for two years until the Army sold them at auction in October 1948. The Walnut Avenue USO club closed on Sunday, May 26, 1946. A year later, on August 5, the federal government sold the building to the city of Lompoc for $20,300. The city paid $4,000 up front and signed a 10-year lease to pay off the balance. The city's Recreation Commission reopened the building as the Lompoc Community Center on September 6, 1947.[153]

## Entertaining the Troops at Cooke During the Korean War

The North Korean invasion of South Korea in June 1950 led to U.S. intervention and the reopening of Camp Cooke two months later to train soldiers for the Korean War. By the time the first soldiers from the 40th Infantry Division began arriving at Cooke in early September, Area Service Unit 6014, as the camp "housekeepers," was already preparing the two service clubs and the Sports Arena for reuse as entertainment centers. The communities of Lompoc and Santa Maria were also gearing up to welcome servicemen. In Lompoc, the Community Center on Walnut Avenue, built originally for the USO during World War II, was renamed the Community and Servicemen's Center. The city's Recreation Commission operated the facility, with Ted Grant as club director. Grant was a former soldier at Cooke during World War II.[154]

The Lompoc chapter of the American Women's Voluntary Service (AWVS), which assisted the USO at the Walnut Avenue clubhouse during World War II, resumed their work at the Community and Servicemen's Center during the Korean War. They were effective agents in recruiting junior hostesses over the age of 18 for dances with soldiers, and acted as chaperons. They also helped plan amateur variety shows and all sorts of other activities and festivities. Many of the shows organized through the AWVS were also performed at Cooke's service clubs. Army and civilian bands took turns playing at these events. Faye Porter, who managed the USO club at Surf during World War II and brought countless shows and activities to the camp, emerged as the entertainment chairperson for the Lompoc chapter of the AWVS. With the assistance of the AWVS, the Community and Servicemen's Center provided military members with a full range of programs and services comparable to any USO club.[155]

In February 1951 the city of Santa Maria donated the building at 410 South Broadway to the USO for use as a service club. This was the same building the USO used during World War II. Around March 1951 the facility reopened under the auspices of the YMCA, with Sam Chollar as club director.[156]

On September 12, 1950, the first of many fine civilian stage repertoires was held at Cooke at Service Club No. 1. Faye Porter organized the variety show of amateur singers and dancers. She would continue to bring additional entertainment to Cooke and the Lompoc Community and Servicemen's Center up to the time the camp closed in 1953.[157]

**An undated image of actress Dorothy Lamour with a group of 40th Infantry Division soldiers at Camp Cooke. Courtesy the California State Military Museum.**

The popular Ralph Edwards radio comedy quiz show *Truth or Consequences* came to Cooke's Sports Arena on September 22, 1950. Soldiers were picked out of the audience to participate in the show's wacky antics.[158]

On November 8, 1950, an all-star musical production direct from Hollywood was staged at the camp Sports Arena. Participating in the Metro-Goldwyn-Mayer extravaganza were actors James Whitmore, Janet Leigh, Keenan Wynn, Marshall Thompson, Arlene Dahl, Vera Ellen, and Arthur Lowe, Jr.; singers Debbie Reynolds and Kay Brown; and pianist, composer, and conductor André Previn, who, incidentally, was in the Army during this time at Camp Cooke.[159]

Marjorie Hall, who gave so much of her talent and time staging dance recitals and teaching dance lessons at Cooke during World War II, presented a program of dance with several of her students at Service Club No. 2 on January 9, 1951. Two days later a group of young Hollywood actors from the Players Ring staged at Cooke the musical comedy *I Love Lydia*.[160]

Popular radio celebrity Bob Hawk brought his radio game show to Cooke on February 20, 1951. Soldiers chosen from the audience were asked a series of questions, and each correct answer earned $5 and a carton of cigarettes from the sponsor. Five correct answers and the contestant qualified for a shot at the grand "Lemac" prize worth $500. Lemac spelled backward was the name of the sponsor.[161]

On March 11, singer and actor Phil Regan aired his Armed Forces radio program nationwide from Camp Cooke. The mostly musical show originated each week from a different military installation or defense plant throughout the United States. Several days later, on March 24, a USO camp show arranged through the Hollywood Coordinating Committee featured Carolina Cotton, the "Yodeling Blonde Bombshell" of Western swing music and B-Western movies; singer Russ Arno; and actors Judy March, Dusty Walker, and Thad Swift.[162]

A second Hollywood extravaganza at the Sports Arena, on March 30, 1951, featured singer and comedian Dennis Day, actress Jean Porter, voice actor and comic Mel Blanc, and a troupe of other entertainers. Blanc was a master of dialect who also did vocal sound effects, and was Warner Brothers' voice for Bugs Bunny, Daffy Duck, Porky Pig, and Tweety Pie.[163]

On April 20 actors Roddy McDowall, Marshall Thompson, Lura Scott, and Bob Peoples, accompanied by music student Jennie Rush, presented lively song and comedy entertainment in two 30-minute shows at the hospital. A week later, on the 27th, USO–Camp Shows presented a fabulous night of entertainment at the Sports Arena featuring the glamorous Ann Sheridan, Sammy Davis, Jr., Eddie Bracken, Dorothy Ford, Maxine Stone, and Danny

**Comedian and entertainer Dennis Day, and camp commander Col. Frank R. Williams. March 30, 1951. Courtesy the California State Military Museum.**

Beck; singers Gale Robbins, Kay Brown, and Jeanne Determann; and Hollywood's honorary mayor, Johnny Grant. Other acts included a vocal combo, a magician, and a comic.[164]

The following month, on May 9, the Hollywood Coordinating Committee featured Broadway showman and radio industry pioneer Nils Thor Granlund; actresses Rita (Kipp) Hamilton, Elinor Donahue, and Diane Jergens; fashion model Jean Smyle; and several other entertainers to a delighted audience at the Sports Arena.[165]

**Actress Jean Porter sings for the boys at Camp Cooke. March 30, 1951. Courtesy the California State Military Museum.**

ANN SHERIDAN  KAY BROWN  JEANNE DETERMANN

**The Ann Sheridan Show filled every seat in the Sports Arena. April 27, 1951. Courtesy the California State Military Museum.**

Fifteen-year-old Elinor Donahue was already a veteran of radio, stage, and movies when she appeared at Camp Cooke. Two years later she would become a beloved television star to millions of viewers as Betty Anderson in the family show *Father Knows Best*. Donohue would later appear in countless other productions, among them *The Andy Griffith Show*, *The Odd Couple*, and *Dr. Quinn, Medicine Woman*.[166]

Asked about her appearance at Camp Cooke, Donahue said that it stands out as one of her most memorable camp shows.[167] She went on to explain that a few weeks before the show was to open at Cooke and at other bases, she received a call from Nils Thor Granlund, inviting her to join the tour. "Everyone," she said, "called him NTG." Donohue was thrilled to be in the group but also realized she would need a dance costume.[168] At that time her mother was working two jobs to manage their finances—at the May Company in Wilshire as a gift wrapper by day, and by night as a seamstress in downtown Los Angeles for a professional costumer, "Christine." Donohue gratefully remembers her mother sewing together the dress that Christine had designed for the show. Donahue recalls:

[The finished dress] was without doubt the ugliest costume I'd ever seen. It was made of magenta taffeta with a pink net ruffle trim. I looked like a cross between a ballerina and a carhop at a 1950s drive-in restaurant. When we went to NTG's rehearsal studio, he absolutely rejected it. But short on money and time, there was nothing to be done. Either wear it or don't go on tour. NTG wanted me in the show, so he turned part of my act into a comedy.[169]

I was to sing a chorus of the Hank Williams song "Anytime," and then do a chorus followed by a half tap dance. NTG said, "Whatever happens while you're singing just keep going. Pay no attention to me." What he did was to return to the stage after my introduction. After I'd sung only a couple of bars, he made a terrible face, rolled his eyes, held his nose, and looked at the audience with disgust at this no-talent gawky kid. It got good laughs, and when I went into my tap routine (and I was a good dancer), we had the audience in the palm of our hands. It was a matinee performance with a full, enthusiastic, and young audience. Just boys, really.[170]

That afternoon at the second show, he was feeling his oats and must have gone overboard because the young men in the audience yelled back at him things like, "Leave the kid alone," "She's doing fine," and so on. That made me feel good because, truth to tell, it had sort of hurt my 15-year-old feelings to be the butt of the joke, having so little confidence in my singing.[171]

Having told this story, I must say that NTG was one of the kindest and most generous men I ever worked for as a young woman. And he was a gentleman.[172]

Playing to a full house of cheering soldiers at the Sports Arena on June 1, 1951, were singers Kay Brown, Maxine Gates, Adele Francis, and Bill White; actors Marvelle Andre, Dorothy Ford, and Danny Beck; harmonica player Keith Thompson; the Knightengales, a musical quartet of two men and two women; and the comedy of Vera Ferguson and Carol Landley.[173]

On June 22, famed ventriloquist Edgar Bergen and his wooden sidekicks Charlie McCarthy, Motimer Snerd, and Effie Klinker drew riotous applauses from soldiers in two shows at the Sports Arena. Joining the performance were vocalist Frances Bergen, wife of the ventriloquist; magician Russell Swan; aerobatic dancer Meribeth Olds; and the comedy act of Mac, Russ, and Owens. The show was arranged by the Coca-Cola Bottling Company, which sponsored Bergen's radio and TV shows.[174]

Less than a month later, on July 17, soldiers and civilians again packed the Sports Arena to attend a performance of *Roadside*, staged by the Pasadena Playhouse troupe of Southern California. The comedy was written by Lynn Riggs, author of *Green Grow the Lilacs*, on which the hit musical *Oklahoma* was based. Later that month, on July 28, the Circle Theater troupe of Hollywood presented a variety show at the Sports Arena. It featured three short one-act comedy skits, and three separate performances by vocalist Eleanor Adler, acrobatic dancer Jerrie Lee Cole, and pianist Bill Howe.[175]

**Bob Hope (right) at Camp Cooke. April 1952. Courtesy the California State Military Museum.**

On August 9, 1951, a troupe of Hollywood entertainers, with actor Marvin Miller, child star Shari Robinson, songstress Pat Michaels, a chorus line of six precision dancers called the "Debbettes," and violinist Jack Jecker, performed at the Sports Arena.[176]

Making his fourth appearance at Camp Cooke, Bob Hope put on two complete comedy shows at Theater No. 3 on April 10, 1952. Among his cast of regulars were Jerry Colonna, Les Brown and his Band of Renown, and Hy Averback, announcer and straight man.[177]

To celebrate the second anniversary of the reactivation of Camp Cooke, a gala ball for soldiers was held at the Sports Arena on August 2, 1952. USO hostesses from Santa Barbara and San Luis Obispo Counties, and from as far away as Los Angeles, were invited to the party as dance partners for soldiers attending the celebration. Ray Herbeck and his orchestra, with vocalist Lorraine Benson, provided the musical entertainment.[178]

Later in August, singer Debbie Reynolds returned to the Sports Arena to thunderous applause. The show included other songstresses, dancers, and jokesters.[179]

On September 16, 1952, actor and musician Buddy Rogers graced the main stage at Cooke. He returned four days later with bandleader and musical composer Sonny Burke and his orchestra for an evening of dance. USO hostesses from Los Angeles and young ladies from the nearby communities were invited to attend the event.[180]

Camp Cooke's mammoth Welfare Carnival and Exposition opened as a four-day festival free to the public on October 30, 1952. Located at Washington Avenue and California Boulevard, the midway included a big top, a Ferris wheel with eleven additional rides for adults and children, and numerous concession booths. Actor Raymond Burr appeared on the final day to give away a 1952 Willys Aero Ace automobile to a lucky ticket holder. Pan American Shows, producer of community fairs in central and southern California, provided many of the attractions. Veteran show promoter Joe Archer donated his services to the Army to help promote the event to raise funds needed for recreational equipment at the camp.[181]

On November 30, 1952, Harry Owens and his Royal Hawaiians band, with singer, musician, and comedienne Hilo Hattie, performed at Camp Cooke's Sports Arena, delighting an audience of about 1,500 soldiers and civilian employees with Hawaiian melodies.[182]

In November the Army began winding down Camp Cooke, and four months later closed the camp for the last time. Entertainment shows continued into early 1953, but with fewer soldiers, larger performances were probably scaled back. In Lompoc, the Community and Servicemen's Center continued to operate as a community resource after the camp closed. Today it is called the Jack Anderson Recreation Center. The USO in Santa Maria returned to its previous incarnation as a youth recreation center. In 1969 the beautiful classical revival building was torn down.[183]

# 6

# THE ARMY DEPARTS

At the turn of the twentieth century, Union Oil of California purchased the subsurface mineral rights to almost half of the land that would later make up Camp Cooke. When the Army acquired this property and other tracts of land, the subsurface mineral rights were retained by the original owners or their assignees. Union Oil was the largest single holder of these rights.[1]

In 1948, Union Oil was permitted to drill two wells in the Rancho de Jesus Maria area of the camp, and two more in the same location in 1952. The first well operated for ten weeks between September and November. It produced about 1,616 barrels of oil but was diluted with more than 10,000 barrels of salt water. It was plugged in June 1949. The second and third wells failed to produce. The fourth well struck oil in September 1952 and indicated a potential for commercial production. A short time later the Army granted Union Oil permission to drill on 1,300 additional acres of land in the Jesus Maria field.[2]

What happened next became a public relations nightmare for the U.S. Army. On November 7, 1952, the Army issued a press release that it was planning to close Camp Cooke probably within the next month to give Union Oil Company unrestricted access to mineral deposits below Camp Cooke. It also indicated that the 44th Infantry Division would be moved to Fort Lewis, Washington. The announcement came as a complete surprise to the residents of Lompoc and Santa Maria, and set off a firestorm of criticism directed at the Army. For the two years that the camp was in full operation, part of this time it supported up to 20,000 servicemen and women, and some 1,200 civilian employees. A large chunk of the camp's monthly payroll of approximately $2 million circulated in Santa Barbara County. Fearing the economic impact on the area if the camp closed, town leaders and business merchants formed committees to pressure their congressional representatives, the governor, and even vice president–elect Richard Nixon to investigate the Army's decision and to have the inactivation order rescinded. They sent committee members to Washington, D.C. and to Sixth Army Headquarters in an effort to keep the camp open. A steady stream of newspaper articles kept the communities informed about the situation and fanned the fires of indignation.[3]

Realizing that it had blundered into a quagmire, the Army issued a second release a few days later that explained it was closing the camp as a cost-saving measure. In addition to Camp Cooke, Camp McCoy, Wisconsin; Camp Drum, New York; and Camp Edwards, Massachusetts, were also scheduled to close. According to the new press release, the inactivation of Camp Cooke would save the federal government about $2 million in the first year, and $4.3 million annually thereafter. The Army also addressed the claim made by the committee organized to protest the inactivation of the camp that $20 million had been spent rehabilitating the installation since its reactivation in 1950. The Army insisted the actual amount was $8.5 million.[4]

Meanwhile, Reese H. Taylor, president of Union Oil, issued a statement intended to clarify the company's relationship with the Army, and to dispel any impression that it had a role in the decision to place the camp on inactive status. Taylor said the firm made no demands on the government, and that development of the company's oil rights at Camp Cooke could proceed without interfering with the military mission.[5]

Whether Taylor was correct we'll never know. After Camp Cooke had closed, Union Oil drilled seven additional wells in the Jesus Maria field between April 1953 and December 1955 before finally shutting down all production in late 1957. Two wells never produced and were capped. The remaining five wells, and No. 4, produced on an irregular basis. The total production history for six of the seven recorded wells including No. 4 was 146,053 barrels of oil and 295,280 barrels of water. About 60 percent of the aggregate came from No. 4. When the Air Force acquired Camp Cooke in 1957, it essentially eliminated the oil encroachment problem by signing memorandums of agreement and making payments in exchange for subordinating the mineral rights.[6]

The committee that opposed the closing of Cooke in 1952 lost credibility when it claimed that moving the 44th Infantry Division was a political patronage deal made by the Democratic administration. Committee members said it was a repayment to the people of Washington for electing Democratic candidate Henry M. Jackson over incumbent Republican Senator Harry P. Cain. By the end of November, organized opposition to closing the camp had ceased to exist.[7]

And so in 1952 Camp Cooke began winding down for a second time. On November 26 an advanced convoy of 44th Infantry Division troops drove out of the camp in the pre-dawn darkness, bound for Fort Lewis. The final contingent of 1,500 division troops left Cooke on December 11. They boarded the U.S. naval transport *Gen. Nelson M. Walker* at Port Hueneme, while long-shoremen and riggers stowed thousands of tons of military vehicles and infan-try equipment aboard four Navy cargo ships. When the ships unloaded at Tacoma, Washington, the troops and cargo went overland to Fort Lewis.[8]

On November 27, 1952, the hospital at Camp Cooke began evacuating patients to Madigan Army Hospital at Fort Lewis. On March 1 the hospital at Cooke closed. The *Cooke Clarion*, which had informed and entertained its readers since its resurgence in September 1950, ended publication on Decem-ber 17, 1952. Three days later the Army auctioned off office and mess hall equipment, furniture, and fixtures having a total value of $10,000. Subsequent auctions also included surplus buildings.[9]

The closing of shops, offices, and services at Camp Cooke coincided with the steady exodus of troops from the camp. In December 1952 the 466th Anti-aircraft Artillery Battalion entrained to March Field near Riverside, California. The 375th Military Police Company had nearly completed its withdrawal dur-ing the final days of December, sending soldiers to Camp Stoneman at Pitts-burg, California, and to other military police units that would serve in Korea. About a dozen police remained behind during the final weeks the camp remained opened. On January 2 the 393rd Ordnance Battalion transferred to Camp Roberts, north of Paso Robles, California. Three days later the 747th Amphibious Tank and Tractor Battalion relocated to Fort Ord at Monterey Bay, California. By this time only the most essential personnel of Area Service Unit 6014 remained on the post to close out the camp.[10]

The end came succinctly and with military precision. Just before sunset at 5:00 p.m. on March 31, 1953, the American flag was lowered over Camp Cooke for the last time. Military staff and a small group of civilians stood at attention and watched as Col. Alexander G. Kirby, commanding officer of the camp, transferred the headquarters flag and command of Camp Cooke to Col. Benjamin B. Albert, commandant of the Branch United States Disciplinary Barracks. Now, for a second time in the installation's history, USDB would be in charge of Camp Cooke and central to all activities on the installation. At the start of the ceremony the prisoner band from the disciplinary barracks sounded Retreat. They played "The Star-Spangled Banner" as the flag was lowered and broke into "Auld Lang Syne" as it was handed off.[11]

To watch over the camp, the Army kept open the post engineer's office, a fire station, and one bachelor officers' quarters to support a caretaker con-tingent of up to 130 security and maintenance personnel.[12]

Shortly before the camp closed, the Army's District Engineer Office in Los Angeles once again began leasing large sections of the installation for grazing and agriculture. By April 1, 1953, almost all of Cooke, excluding the area around the disciplinary barracks, had been leased. Former leaseholders, who were forced to leave the reservation when Cooke reopened in August 1950, were given first choice to return to the camp and finish out their leases when the Army announced in November 1952 that it was closing the camp. Other ranchers and farmers also received the opportunity to submit bids. The leases were valid through November 1953. Beginning in October 1953, the Army solicited new bids for a five-year period.[13]

## The U.S. Branch Disciplinary Barracks

One of the more notable events at the disciplinary barracks occurred in July 1953 when, after some four months of confinement, thirty-seven Puerto Rican soldiers were released and restored to full active duty. They were part of a larger group of 103 Puerto Rican enlisted members of the 65th Infantry Regiment charged with disobeying the lawful orders of an officer while on duty in Korea, misbehavior before the enemy, and desertion. The incidents happened at two different locations between October and November 1952. Multiple courts-martial for groups of soldiers ended with five defendants being acquitted and eight others having charges against them dismissed. Ninety-one soldiers, including one junior officer, received sentences with varying degrees of punishment.[14]

New construction at USDB began in late 1954 on several permanent facilities for staff personnel. The first of these projects consisted of two three-storey barracks to house 526 soldiers, and a connecting mess hall building. The Army's District Engineer Office in Los Angeles awarded the project to the Edward R. Siple Construction Company in October. At the dedication ceremony for the mess hall on January 20, 1955, the keynote speaker was Maj. Gen. William F. Dean, deputy commander of the Sixth Army and recipient of the Medal of Honor, earned as a prisoner of war in Korea. The two barracks were completed four months later and replaced the temporary quarters which were moved from Camp Cooke when the USDB was first activated. That same year the main road between the prison and Santa Lucia Canyon Road was named Klein Boulevard in honor of Lt. Col. Raymond E. Klein, who was the deputy commandant in charge of operations at USDB. Klein died at his home in Lompoc on September 23, 1955. Born in Brooklyn, New York, the 45-year-old Klein served in the European Theater of Operations during World War II.[15]

Between 1956 and 1957, several additional construction projects were initiated on the installation, beginning with family housing that went up across from the prison on the east side of Santa Lucia Canyon Road. E. H. Moore & Sons, of San Francisco, built 36 two- and three-bedroom units in January 1957. Two years later additional homes were constructed behind the first 36 units. In July 1956 the Lake Canyon recreational area opened at Camp Cooke for USDB staff. It featured an artificial lake approximately 100 yards long and stocked with warm-water fish. A year later, in July 1957, a red brick chapel with a seating capacity of 175 opened at USDB near the new mess hall. The last major construction project at USDB during this period was a 10,000-square-foot gymnasium. The project began in February 1957 and was completed around the end of the year.[16]

By the mid–1950s the inmate population at USDBs across the United States, including the one at Camp Cooke, was on a downward trend. No longer did the Army require the prison at Cooke, and on August 1, 1959, it transferred the facility to the U.S. Bureau of Prisoners to house civilian inmates. The outgoing USDB commandant, Col. Weldon W. Cox, was succeeded by Warden Grieg V. Richardson. Cox was reassigned to Fort Leavenworth, Kansas. During its twelve years of operation, the USDB had processed more than 20,600 prisoners. Immediately following its transfer, the prison was renamed the Federal Correctional Institution.[17]

# 7

# A NEW MISSION

Toward the end of World War II, Nazi Germany developed the V-2 rocket as a new weapon to use against the western Allies. Packed with a ton of high explosives and with a range of more than two-hundred miles, the first of these flying bombs slammed into Paris on the morning of September 8, 1944. Later that evening additional missiles fell on London, marking the start of a terrifying offensive that would include attacks on other targets in Europe. Formidable as these weapons were, if introduced before D-Day they still would not have altered the outcome of the war.[1]

As the war ended, America and the Soviet Union hastened to capture V-2 experts, plans, and rockets, and to use these assets to jump start their own respective missile programs. In 1946 the U.S. Air Force began to subsidize a research program by the Convair Aircraft Corporation for a long-range missile, but the funds were cut off the following year. Efforts to develop an intercontinental ballistic missile (ICBM) program were stymied by austere defense budgets and competition from proven deterrents to aggression—long-range bombers. This left American missile development to concentrate on space probes using modified V-2s and newer short-range rockets, many of which were battlefield weapons.[2]

During 1950, new scientific studies indicated that it was technically feasible to build long-range missiles, which led to the revival of the Convair project on a small scale in January 1951. Still unsettled were the size requirements of the missile, the amount of thrust needed to carry an atomic warhead, and a plan for achieving a high level of target accuracy. An answer to the last of these questions came in November 1952 when the first hydrogen bomb was detonated at the Eniwetok Atoll, a Pacific island chain some 2,380 miles southwest of Honolulu. The vast destructiveness of the bomb meant that accuracy requirements could be relaxed. In March 1954, thermonuclear tests confirmed the feasibility of developing lighter and more powerful warheads. As a result,

Convair's Atlas missile design configuration could be significantly downsized, and the stringent 1,500-foot accuracy requirement reduced to between two and three nautical miles. This was loosened to five miles after the Atomic Energy Commission predicted it could develop a one-megaton warhead light enough to be carried on a trimmed-down Atlas. The breakthrough technology came at a time of growing concerns about the Soviet Union's missile development and its successful testing of a hydrogen bomb in August 1953.[3]

In the months to follow, the Air Force selected the Martin Company to develop the Titan I ICBM as an alternate to Convair's Atlas. Added to the mix was Douglas Aircraft Company's Thor intermediate range ballistic missile (IRBM). The Thor grew out of a meeting of the government's Scientific Advisory Committee in January 1955. In February the United Kingdom expressed an interest in the Thor. Since an American IRBM could be made available earlier than an ICBM, basing the 1,500-mile-range missile in the U.K. would be a short-term deterrent to Soviet aggression in the region. President Dwight D. Eisenhower agreed to the request and on December 1, 1955, approved a National Security Council recommendation assigning the Thor "joint" highest national priority with the Atlas and Titan.[4]

Support for the missile program in the United States was picking up momentum. On July 1, 1954, the Air Force's Air Research and Development Command established the Western Development Division to supervise ballistic missile development. Headquartered at Inglewood, California, the WDD became the Air Force Ballistic Missile Division on June 1, 1957, and by that time had added reconnaissance satellite programs to its management role, though none had yet been launched. In order to speed up progress in its ballistic programs, WDD adopted the concept of parallel or concurrent development. All of the elements that went into the weapon system—development, testing, facility construction, site installation, and training—all proceeded simultaneously or at least in a very narrow and overlapping timeframe.[5]

## *The Air Force Takes Charge*

With the advent of the missile age in the 1950s, the Air Force built the first ICBM and IRBM test facilities in Florida at Cape Canaveral near Patrick Air Force Base. The launch complexes were oriented toward research and development testing and did not resemble hardened operational complexes that would be needed at field bases. In January 1956 the Air Force embarked on a search for a suitable launch site to conduct initial operational testing of ballistic missiles and the training of missile launch crews. The site would also serve as

America's first combat-ready ICBM missile base equipped with nuclear warheads. A special site selection committee was formed that examined more than two-hundred potential sites before recommending the vacant Camp Cooke in June 1956. Cooke stood out from the other sites for several reasons. First, it was large and remote, making it easy to disperse many missile launch and support facilities. This same feature ensured that the launchers would be a safe distance from civilian communities and the base's cantonment or living area. A large number of existing buildings at the camp would provide housing for personnel and for other support functions. But most importantly, its coastal location would allow missiles to be flown over water without risk to populated areas.[6]

The recommendation was passed to the Secretary of the Air Force and approved by Secretary of Defense Charles E. Wilson. On November 16, Wilson directed the Army to transfer slightly over 64,000 acres of northern Camp Cooke to the Air Force, excluding the area around the Branch United States Disciplinary Barracks.[7]

On January 25, 1957, the Army issued the Air Force a temporary use permit for Camp Cooke north of the Santa Ynez River. By this time, the Air Force presence at Cooke consisted of Maj. Frederick K. Smith, an installations engineer, and Capt. Willis G. Shaneyfelt. TSgt. Robert D. Roller, a telephone communications expert, arrived at the camp on February 24. The scene that met these men consisted of a cluttered mass of dilapidated World War II buildings amid weeds and brush growing everywhere. Roads were in need of extensive repair, and much of the telephone system was inoperative.[8]

To operate the missile base, Air Research and Development Command activated Headquarters, 392nd Air Base Group at Cooke, and the 1st Missile Division at WDD in Inglewood on April 15, 1957. The Air Force contingent at Cooke numbered about five officers and fifteen airmen. With the activation of the 704th Strategic Missile Wing (Atlas) at Cooke on July 1, the 392nd was assigned to the wing. On July 16 the 1st Missile Division, with Col. William A. Sheppard in command, relocated to Cooke to supervise wing operations. During this formative period the work of these latter two organizations involved planning for missile operations and training. The division was assigned to the newly renamed Air Force Ballistic Missile Division in Inglewood.[9]

Meanwhile, on May 9 a small group of newsmen and local dignitaries gathered in the shade of a eucalyptus windbreak to witness formal groundbreaking ceremonies presided over by Maj. Gen. Osmond J. Ritland, vice commander of Western Development Division. Gen. Ritland spoke about the vast construction activities that would follow and the work already in progress.[10]

In preparation for its new mission, the former army camp underwent a parallel construction and renovation program. The Air Force's construction

The main cantonment area of Camp Cooke in early 1957. U.S. Air Force.

agent, the Army Corps of Engineers, hired and supervised civilian contractors to perform the work. Construction started in late April 1957 when the P. J. Walker Company of Los Angeles moved in employees and heavy equipment to begin work on the Atlas guidance station. All through the summer and fall of 1957 construction crews built technical launch and support facilities that included missile assembly buildings, tracking stations, operations centers, and launch complexes. On June 13, 1957, Fredericksen, Kasler, and Stolte, Inc., began building the first Atlas launch facility. Progressing simultaneously on renovating support facilities, and repairing roads and utilities was the George A. Fuller Company. The entire camp was undergoing a revitalization, all part of the tremendous task of transitioning Camp Cooke from an abandoned Army post into a modern missile base, due largely to the influence of the V-2 rocket. On June 7, 1957, the Air Force renamed its part of the installation Cooke Air Force Base, and exactly two weeks later the departmental transfer of the camp to the Air Force was finalized.[11]

One of the earliest problems at Cooke AFB was a shortage of housing for the growing population of airmen, civilians, and contractor personnel. With military family housing non-existent on base, and too few rental vacancies in Lompoc and Santa Maria, rents on available properties skyrocketed as the competition for housing increased. The housing situation was particularly acute for lower ranking airmen with families. Some resorted to living in sub-standard housing, while others commuted up to sixty miles from such distant communities as Santa Barbara, Pismo Beach, and San Luis Obispo. Realizing that on-base housing was essential, the Air Force obtained congressional approval for construction of new three- and four-bedroom Capehart family housing. In July 1957 contractors began moving or demolishing the old wooden buildings in the cantonment's northwest corner, site of the former German prisoners of war camp, and on October 23, 1957, the groundbreaking for the first increment of 880 Capehart homes was held. Gradually, long rows of modern ranch-style homes took shape on streets pleasantly named for trees. Additional increments followed over the next several years, eventually bringing the number of homes on-base to more than 2,000.[12]

Meanwhile, initial increments in the Capehart project did not eliminate the critical housing shortage. To further ease the problem, trailer court sites were set aside on base. And for unmarried airmen, mobilization-type barracks were converted into modern dormitories. Bachelor officer quarters were also converted into spacious apartment units. By the fall of 1957 the Air Force was well on its way toward filling its ranks at Cooke and dotting the landscape with new and refurbished facilities.[13]

On October 4, 1957, the Soviet Union surprised the world with the launch of its Sputnik earth-orbiting satellite, and repeated this achievement a month later with Sputnik 2, which carried a dog into space. The message drawn from these launches was that rockets, which placed satellites overhead, were also capable of dropping atomic bombs on the United States. By launching the first Sputnik, the Soviets had set a precedent for the principle of free flight in space regardless of national boundaries.[14]

Just as the Army had hastened to ready Camp Cooke after the Japanese attack on Pearl Harbor, the Air Force response to the Soviet challenge was to immediately accelerate development of its missile program. It also transferred management responsibilities for Cooke AFB from Air Research and Development Command to the Strategic Air Command on January 1, 1958. Along with the transfer, SAC acquired the three ARDC base organizations (1st Missile Division, 704th Strategic Missile Wing, and 392nd Air Base Group), and responsibility for attaining initial operational capability (IOC) for the nascent missile force. The SAC mission would also include training missile launch crews and obtaining Emergency War Order capability (later called Strategic Alert) in which intercontinental ballistic missiles equipped with nuclear warheads would be ready to retaliate if the United States were attacked.[15]

ARDC retained responsibility for the design and activation of launch and support facilities. It also retained research and development testing of ballistic missiles and space systems. These activities were carried out by a Ballistic Missile Division Field Office established at Cooke immediately after the realignment. When developmental testing was completed on each missile system, the vehicle program was handed off to SAC for operational testing and training exercises. Space launches were conducted by ARDC and SAC units, with the vast majority of these operations handled by ARDC.[16]

On October 4, 1958, exactly one year after the first Sputnik launch, Cooke Air Force Base was renamed Vandenberg Air Force Base in honor of Gen. Hoyt S. Vandenberg, the second Chief of Staff of the Air Force. On the following day, Mrs. Vandenberg and their son, daughter-in-law, and grandson were the guests of honor at formal dedication ceremonies held at the base parade grounds, attended by several hundred guests. Mrs. Vandenberg arrived in a 1958 pink Cadillac with huge tailfins. Among the speakers on the podium that day were Maj. Gen. David Wade, the new 1st Missile Division commander, who commented on the proud tradition that the base was inheriting.[17]

## The Navy at South Camp Cooke

Some one-hundred miles south of Camp Cooke, at Point Mugu, the U.S. Navy had been operating a small offshore guided missile range since 1946. In

1956 the Navy and the Air Force indicated to the Department of Defense interest in acquiring Camp Cooke for their respective missile programs. The Air Force quickly secured its requested section of the camp in November 1956. The Navy submitted its request in April 1957 to the Special Committee on Adequacy of Range Facilities (SCARF). The SCARF was established several months earlier by the Secretary of Defense to examine missile ranges in the United States. In September the committee recommended the establishment of a national Pacific Missile Range. It also recommended the Navy be in charge of the range, and that it receive Camp Cooke south of the Santa Ynez River.[18]

The new Secretary of Defense, Neil H. McElroy, approved the SCARF recommendations, and on December 7, 1957, shared his decision with the Secretary of the Navy. Following a now familiar path, the Army granted the Navy a temporary permit to Camp Cooke on February 14, 1958. Later that month the U.S. Coast Guard transferred its rescue station near Point Arguello to the Navy. On May 10 the Navy commissioned its part of the former Army camp as Naval Missile Facility Point Arguello. The conveyance of all 19,861 acres from the Army to the Navy was finalized on May 27, 1958. Over the next six years the Navy built launch and tracking facilities at Point Arguello, and established a network of missile instrumentation sites that extended far into the South Pacific area.[19]

Recognizing that launch missions from Cooke and the operation of the Pacific Missile Range (PMR) were clearly related, the Air Force and the Navy entered an agreement for coordinated use of the range. The agreement was signed by Air Force Chief of Staff Thomas D. White and Navy Chief of Operations Arleigh Burke on March 5, 1958. Although the agreement appeared to provide the basis for a harmonious relationship in the Cooke-Arguello area, a bitter fight broke out behind the scenes and in Washington at the highest levels. The most serious disagreements were over space missions, and started in February 1959 when the Air Force began launching Discoverer satellites that flew over the small community of Surf. After the third and fourth missions, launched June 3 and 25 respectively, failed to reach orbit, the Air Force changed the azimuth to an easterly direction beginning with the fifth mission on August 13. The adjustment would make it easier to place the satellite into orbit by taking advantage of the earth's rotation. In doing so, the vehicle charted a path further inland that took it directly over the Navy's Point Arguello facility and to the south at Sudden Ranch. The Navy opposed these missions on the basis they represented an unacceptable risk to citizens and private property should the rocket fail during liftoff. The Air Force contended the risk was minimal. The Navy countered by citing certain paragraphs of the Burke-White Agreement that gave final authority for missile flight safety to

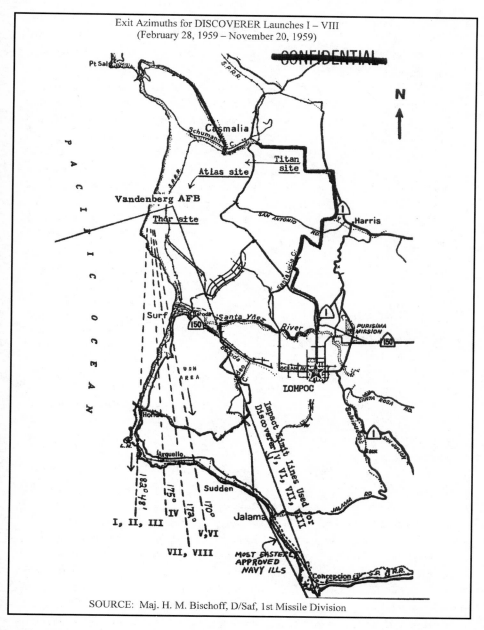

Exit Azimuths for DISCOVERER Launches I – VIII
(February 28, 1959 – November 20, 1959)

SOURCE: Maj. H. M. Bischoff, D/Saf, 1st Missile Division

**The Air Force moved the launch azimuths easterly by several degrees in an effort to make it easier to place a satellite into orbit during the early Discoverer space launches. This resulted in Air Force rockets launched from Vandenberg flying directly over the U.S. Navy's Point Arguello facility and further south at Sudden Ranch. U.S. Air Force.**

the PMR Commander. Although the Navy acceded to the Discoverer flight plans, it halted all traffic on the main line of the Southern Pacific Rail Road that passed through the former Camp Cooke, and continued to evacuate Surf, with its population of less than fifty people, prior to each Discoverer launch.[20]

Such disagreements were only part of a more deeply rooted mistrust at PMR between the Navy and Air Force that grew from the Department of Defense's failure to clearly define the roles and missions of the services, and the absence of a centralized control within the department over range development and use. The Air Force carped that the Navy was trying to use PMR to create its own space program at the Air Force's expense. The Navy responded that it was only trying to develop PMR according to its charter as a national launch range for the Navy and other users. It accused the Air Force of seeking to block further development of PMR in order to make Vandenberg AFB the center for research and development of military space programs on the West Coast. The Navy was, in fact, looking to expand the PMR scope beyond the original SCARF concept to include establishing its own satellite base at Arguello. Feeding into this maelstrom of distrust was the Advanced Research Projects Agency. Established in February 1958 as a Department of Defense entity, ARPA's role was to create new technologies and divide up military satellite programs between the Services.[21]

Confronted by range management issues at PMR and at other missile ranges, Secretary of Defense McElroy appointed an advisory committee to examine the issues and offer recommendations. Under the chairmanship of Walker L. Cisler, president of the Detroit Edison Company, the committee submitted its findings on November 30, 1959. Among its recommendations was the establishment of a central control agency at the Department of Defense level overseeing range development and use. The office was created in May 1960 as the Office of Range and Space Ground Support within the Directorate of Defense Research and Engineering. This high-level authority resolved many of the problems festering at Arguello/Vandenberg, and at other missile ranges.[22]

During this time, Admiral Burke and the Air Force Vice Chief of Staff Curtis LeMay signed a new accord relating to Arguello/Vandenberg on September 22, 1959, that replaced the March 1958 agreement. It gave the Navy full responsibility for missile flight safety (except for ballistic missile training launches conducted from Vandenberg). To assuage Air Force concerns about the Navy trying to take over its space role, the Air Force was permitted to retain full control over the systems it was developing and to continue launching satellites from Vandenberg. The agreement came four days after Secretary McElroy issued his assignment of service responsibilities in which the Air Force received the most prominent role in space booster and satellite systems. On December

31, 1960, an amendment to the Arguello/Vandenberg agreement expanded PMR's destruct authority to include ballistic missile training launches from Vandenberg.[23]

On November 16, 1963, Secretary of Defense Robert S. McNamara directed sweeping management changes to the Department of Defense's ballistic missile and test range facilities across the country as part of an overall cost-savings consolidation effort between the military services. Among the targets, the Pacific Missile Range was to be turned over to the Air Force in two parts. The transfer began on July 1, 1964, with the decommissioning of Naval Missile Facility Point Arguello, and the entire property annexed to Vandenberg. The transfer was finalized on February 1, 1965, when all range support for ICBM and space launches from Vandenberg, as well as many of the Navy's instrumentation tracking sites in the Pacific, also transferred to the Air Force. Managed as a national range under Air Force control, the former PMR was renamed Air Force Western Test Range and is known today as the Western Range.[24]

## Additional Land for a New Space Program

Six months after the transfer of the Pacific Missile Range to the Air Force, President Lyndon Johnson announced on December 25, 1965, that development work would begin at Vandenberg AFB on a new space program called the Manned Orbiting Laboratory. When the Defense Department first proposed the MOL program in early 1964, it was to be based at Cape Kennedy in Florida, and its mission was to explore the military usefulness of man in space. The vehicle configuration would consist of a two-man Gemini B spacecraft attached to a laboratory vehicle, boosted into space by a Titan IIIC. (Increases in the weight of the laboratory vehicle later necessitated a change in the booster to a Titan IIIM.) Once in space, the astronauts would transfer from the Gemini capsule to the laboratory and for the next thirty days conduct scientific experiments that would have military applications. At the end of the mission the crew would return to Earth in the Gemini, leaving behind the laboratory to de-orbit and burn up on reentering the atmosphere.[25]

Some critics believed that compared to NASA's Mercury and Gemini space programs at the Cape, the MOL's scientific experiments didn't justify its proposed hefty cost. After extended military study and congressional inquiry, the MOL was moved to Vandenberg. Launched into polar orbit, its primary mission would be manned space reconnaissance of the Soviet Union. Despite the mission changes, there was no public disclosure about the true

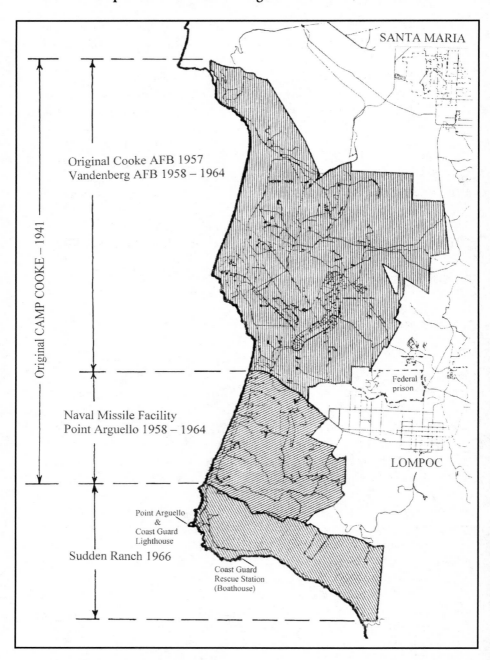

SANTA MARIA

Original Cooke AFB 1957
Vandenberg AFB 1958 – 1964

Original CAMP COOKE – 1941

Federal
prison

Naval Missile Facility
Point Arguello 1958 – 1964

LOMPOC

Point Arguello
&
Coast Guard
Lighthouse

Sudden Ranch 1966

Coast Guard
Rescue Station
(Boathouse)

**Since 1941, the military installation grew in increments to its present size of 99,099 acres. U.S. Air Force.**

plans for MOL from either the President or in the congressional hearings held in February 1966.[26]

At Vandenberg, the Air Force would build a new launch complex for MOL called Space Launch Complex 6, but it first had to acquire the property south of the base which was in private hands. When direct negotiations with the property owner failed to reach a compromise price, the government obtained the land through condemnation proceedings under the power of eminent domain. It filed a Declaration of Taking with the Federal District Court in Los Angeles on February 28, 1966. The court concurred, and the following day 14,903.8 acres were added to Vandenberg AFB. Of this acreage, 14,404.7 was Sudden Ranch (owned by Sudden Estate Company), and 499.1 acres were from the Scolari Ranch. The court ordered that $9,002,500 be paid for the land. The annexation of the property increased the size of the base to its present 99,099 acres. Today, Vandenberg stands as the third largest Air Force Base, after Eglin AFB in Florida and Edwards AFB in California.[27]

Construction of MOL facilities started at the former Sudden Ranch property on March 12, 1966. While this work proceeded reasonably well, other aspects of the program were mired in technical and budget problems that caused repeated delays in the launch schedule. Beginning in 1966 and continuing over the next three years, funds were curtailed to finance the expanding war in Vietnam, President Johnson's "Great Society" programs, and the Apollo Moon project. By 1969 the first MOL flight had slipped from 1968 to late 1970, and the original program cost doubled to at least $3 billion. Externally, the MOL was also coming up against emerging technologies that included concepts for a reusable space launch vehicle that would later evolve into the Space Shuttle. On June 10, 1969, Defense Secretary Melvin R. Laird informed the Senate Appropriations Committee that President Richard Nixon had cancelled the MOL program. Construction work at the launch site was about 90 percent complete. The launch complex was mothballed until it could be used by another program.[28]

## Select Launches from Vandenberg AFB

On a clear Tuesday afternoon on December 16, 1958, some 182 invited members of the press and an almost equal number of dignitaries and other guests filled every seat on specially erected bleachers to watch from a safe distance the first missile launch from Vandenberg AFB, a Thor intermediate range ballistic missile (IRBM). The vehicle was stored horizontally in a two-piece shelter that resembled a Quonset hut on wheels. During the launch

**A Thor IRBM was the first missile launched from Vandenberg AFB. December 16, 1958. U.S. Air Force.**

countdown the shelter separated, exposing the 65-foot-long missile, which was slowly raised by the transporter-erector to a vertical position for fueling and firing. At 3:45 p.m. the single-stage ballistic missile lifted off from Complex 75-1-1. As the missile ascended into the skies, leaving behind a contrail of while smoke, the crowd of spectators broke into cheers. The roar of missile engines could be heard for miles around.[29]

Thor 151 carried an inert warhead 1,268 nautical miles to a broad ocean area target in the Pacific. The actual impact was approximately three nautical miles to the left and six nautical miles over the target center. The eight-man launch crew was supervised by Capt. John C. Bon Tempo. They were assigned to the 392nd Missile Training Squadron of the First Missile Division's 704th Strategic Missile Wing.[30]

As part of an Anglo-American agreement to train and equip four British Royal Air Force squadrons with Thor missiles in England, all twenty-one remaining launches in the program were conducted from Vandenberg by RAF crews training under the supervision of U.S. Air Force instructors from the 392nd Missile Training Squadron. In some instances, graduating classes ended their final session by launching a Thor missile. The first of these operations, nicknamed "Lion's Roar," was successfully conducted on April 16, 1959. To fully test the reliability of the missile system and combat crew proficiency, other British crews returned to Vandenberg to launch Thor missiles that had been based in England. The Thor IRBMs in England were under the operational control of the British government, but U.S. Air Force personnel controlled their thermonuclear warheads. On June 18, 1962, an RAF crew launched the last Thor IRBM. By now, the Thor was an obsolete weapon. In 1963, it was removed from operational service in England and returned to the United States.[31]

By adding an upper (second) stage motor to the Thor IRBM, the Air Force turned it into a space launch vehicle for carrying satellites into orbit. The first of these combinations to be used at Vandenberg, a Thor-Agena A, carried into near polar orbit the Discoverer I spacecraft on February 28, 1959. Shrouded in a cover story of being a scientific program with the innocuous name Discoverer, it was actually America's first covert space reconnaissance program, code-named Corona and under the auspices of the Central Intelligence Agency. Discoverer missions carried Corona or other types of reconnaissance cameras and, after several orbits around the Earth, ejected recoverable canisters containing exposed film. In a perfect scenario the canister would be snagged in midair by C-119 and C-130 aircraft specially equipped with a large trap that extended out the rear of the plane. If the aircraft missed, the canister could be retrieved from the ocean by a recovery team of ships, helicopters, and divers in a technique later employed by NASA during its manned Mercury, Gemini, and Apollo programs.[32] Unlike most Discoverer missions, the first flight carried only engineering equipment to record vehicle performance. To lend credence to the cover, the second vehicle carried a small

**The Discoverer XIII lifted off from Vandenberg on August 10, 1960. The recovery of the Discoverer capsule the following day made it the first man-made orbiting object returned to Earth. U.S. Air Force.**

biomedical experiment into orbit and vanished after it mistakenly came down near Spitzbergen, Norway, close to the Russian border. The third mission on June 3, 1959, was the first and only launch from Vandenberg to carry live cargo—four black mice. Black mice were chosen so that scientists could study the possible hair-bleaching effects of cosmic rays on the recovered mice. Instead of boosting the spacecraft into orbit, the Agena second stage motor apparently fired downward, sending the capsule containing the mice into the Pacific Ocean, or burning up on reentry.[33]

After a string a failures, the first successful recovery of a Discoverer capsule, and the first man-made orbiting object returned to Earth, occurred on the thirteenth mission, on August 11, 1960. The capsule was plucked from the ocean 330 miles northwest of Hawaii. Unfortunately, this was a diagnostic flight without camera or film. Success finally smiled on Discoverer XIV. Launched on August 18, the capsule, containing twenty pounds of exposed film, ejected from the spacecraft on the seventeenth pass and was recovered in mid–air. This one satellite mission provided more photo coverage of the Soviet Union than all the previous U-2 aircraft missions combined. Images collected by Discoverer would debunk the so-called missile gap with the Soviet Union and help shape the course of America's ICBM program. Having worn out its cover story, the Discoverer name was retired in January 1962. The Corona program continued for another decade under the code name KEY-HOLE.[34]

America's first lady of intercontinental ballistic missiles, the Atlas, flew its first West Coast mission from Vandenberg on September 9, 1959. Atlas 12D lifted off from Complex 576A, pad 2 (576A-2), and sent an inert warhead to a water target near Wake Island, 3,899 nautical miles from the launch site. Although scoring accuracy was secondary to demonstrating the capability to launch the missile successfully, the reentry vehicle impacted less than a nautical mile to the right and 1.2 nautical miles short of the target center.[35]

After witnessing the launch, Gen. Thomas S. Power, commander in chief of the Strategic Air Command, declared the Atlas system operational. While additional launches would be needed to fully validate the system, SAC would begin integrating the Atlas into its Emergency War Order plan. One month later on October 31, pad A-1 of three launch pads at Complex 576A became the first American launch site to hold on strategic alert an Atlas missile equipped with a nuclear warhead. Occasionally a second and infrequently a third Atlas managed to be placed on alert at the same time.[36]

This event marked the beginning of a transition period from relying solely on bombers to a mixed strategic force with ballistic missiles as the Air Force's leading long-range strategic weapon delivery system. The change came at a time when advances in the Soviet Union's air defense capabilities were making American long-range bombers increasingly ineffective as a deterrent to a potential Soviet first strike. Indeed, only a few months later on May 1, 1960, the Soviet Union shot down a high-flying American U-2 reconnaissance aircraft over its territory.

**The first Atlas ICBM launch from Vandenberg AFB. September 9, 1959. U.S. Air Force.**

Unlike hardened underground silos built for later ICBMs that protected the vehicle from the harsh outdoor elements of wind, rain, and saltwater mist, and possible enemy actions, the missiles at Complex 576A were stored outdoors in a vertical configuration on a launch mount next to a large gantry tower. To combat the environmental effects on the launch system, technicians periodically removed each missile from the launch mount for routine maintenance and inspection. When they began removing the first missile on December 31, they discovered that the launch mount was frozen to the launch assembly. This situation raised considerable doubt whether a successful launch would have been possible.[37]

The alert posture at 576A ended on February 6, 1961, for several reasons, primarily because the complex was designed as a test facility rather than for real-world operational use. By 1961 other ICBM bases had opened around the country, and, with the latest in hardened shelters, they became the primary locations for the strategic alert mission.[38]

The Titan I was America's second ICBM and was deployed in underground silos. A large elevator at the base of the silo raised the missile to the surface for propellant loading and firing. By the time the Titan I entered testing at Vandenberg in 1961, the more advanced Titan II, with storable propellants, all inertial guidance, and an in-silo launch capability, was already in development. A full-scale test of a Titan missile launched from inside a silo was needed to determine the blast effects on the vehicle and the facility.[39]

The launch test was conducted at Vandenberg using a specially configured Titan I. Its first-stage motor was filled with propellant. The second stage was ballasted with water. A dummy reentry vehicle attached to the upper stage simulated an actual warhead in size, weight, center of gravity, and moment of inertia. The missile stages were then lowered into a prototype Titan II silo called the Silo Launch Test Facility (SLTF). On May 3, 1961, the button was pressed and the missile successfully exited the silo, validating the silo launch

Three Atlas D ICBMs on operational alert at Vandenberg's Launch Complex 576A. U.S. Air Force.

concept. At the conclusion of the flight test, the missile flight safety officer destroyed the vehicle over the ocean about 175 seconds after liftoff. This test marked the first Titan launch from Vandenberg and the nation's first launch of a Titan from an underground silo.[40]

The second and complete Titan I missile system was launched on September 23, 1961, but a premature engine shutdown caused the reentry vehicle to fall 480 nautical miles short of its broad ocean area target. The last Titan I lifted off from Vandenberg on March 5, 1965, racking up a total of twenty launches with mixed results.[41]

The first Titan II missile launched from Vandenberg AFB, and the first nationally from an underground silo, got off to a disappointing start when it exploded 18,000 feet over the Pacific Ocean on February 16, 1963. A second attempt, on April 27, 1963, succeeded in sending its reentry vehicle to a broad ocean area off the coast of Wake Island.[42]

The solid-fueled Minuteman I intercontinental ballistic missile entered the Research and Development program at Vandenberg AFB in 1962. Unfortunately, the first two missiles (launched September 28 and December 10, 1962) broke apart and exploded seconds after exiting their underground silos. The third attempt, on April 11, 1963, performed beautifully. It deployed a single reentry vehicle to a broad ocean area near Wake Island.[43]

Alouette 1, the first international spacecraft launched from Vandenberg AFB, was a joint effort between Canada and the United States. The 320-pound spacecraft was designed and built in Canada, making it the third nation in the world, behind the United States and the Soviet Union, to enter the space age. Its panel of instrumentation collected valuable data about the upper ionosphere. Alouette 1, or Topside Sounder spacecraft, was launched by NASA aboard a Thor-Agena B from Complex 75-1-1 on September 28, 1962, and provided data for the next several years.[44]

On February 24, 1966, a Strategic Air Command missile combat crew from the 341st Strategic Missile Wing, Malmstrom AFB, Montana, fired the first dual, or simultaneous, launch of two Minuteman I intercontinental ballistic missiles from Vandenberg AFB. The missiles lifted off from underground silos, and both successfully deployed a reentry vehicle to the Eniwetok Lagoon. In the second pair of Minuteman launches, on December 22, 1966, a crew from the 455th Strategic Missile Wing, Minot AFB, North Dakota, fired the first missile that sent a reentry vehicle to the Eniwetok Lagoon. The second missile, launched by a crew from the 90th Strategic Missile Wing, F. E. Warren AFB, Wyoming, malfunctioned and was destroyed by the missile flight safety officer fifty-four seconds after liftoff.[45]

Throughout the early 1960s, modified Thor and Atlas missiles fitted with

**The first Titan IIIB from Vandenberg AFB carried a reconnaissance satellite into orbit. July 29, 1966. U.S. Air Force.**

The SV-5D PRIME spacecraft ready to be hoisted atop the Atlas missile in the background. December 1966. U.S. Air Force.

upper-stage motors were launched from Vandenberg AFB. Many of them carried reconnaissance satellites into orbit. The KEYHOLE or KH series of spy satellites had replaced the designation Corona. The third-generation KH-8 was heavier than its predecessors and weighed an estimated 6,600 pounds. To lift this spacecraft the Air Force introduced the Titan IIIB, which was essentially a modified Titan II ICBM with an Agena D upper (third) stage motor. The first Titan IIIB Agena D to carry a KH-8 lifted off from Vandenberg on July 29, 1966.[46]

In the 1960s the Air Force embarked on a series of hypersonic space programs. One of these programs was known as PRIME (Precision Recovery Including Maneuvering Reentry). Its dual objectives were to test new developments in space systems and to explore the possibilities of developing future space vehicles that could maneuver through the atmosphere like an airplane.[47]

PRIME consisted of the SV-5D lifting body vehicle, essentially a wedge-shape wingless aircraft with two stubby fins that gained lift from the flat bottom shape of its fuselage. Relatively small, it was 6.6 feet long, 2.8 feet high, and 4 feet wide at the top of the aft fins. At liftoff the vehicle weighed approximately 895 pounds. According to the flight plan, the SV-5D would be launched from an Atlas SLV-3 booster from Space Launch Complex 3 East at Vandenberg AFB. Following its boost to altitude, the Atlas was to point the wingless vehicle toward earth and release the SV-5D at hypersonic speed. During its simulated trajectory of a spacecraft returning from orbit, the vehicle would use nitrogen gas thrusters for altitude and control in space (like other spacecraft), and hydraulically-actuated flaps (like conventional aircraft) for maneuvering within the atmosphere during reentry. Guidance for most of the mission was to be performed by an onboard inertial system. Terminal guidance to a pre-selected target and recovery area near the Kwajalein Atoll would be provided by a ground-based system.[48]

Following its separation from the Atlas booster at an altitude of about 100,000 feet, the first of two parachutes would deploy to decelerate the vehicle. At 50,000 feet the main chute would deploy, further decelerating the craft to 10,000 feet where it would be snagged and recovered by a JC-130 aircraft. As a backup mode, the vehicle would impact in the ocean and use a floatation system until it could be recovered by surface vessels.[49]

Four flights were planned, with the first of these conducted as a system verification test on December 21, 1966. During this mission the SV-5D separated from its Atlas booster and simulated a reentry from a low-earth orbit with a straight-line entry. Despite an otherwise nominal mission in which the lifting body landed within 900 feet of the pre-selected target point, failure of the main chute to fully deploy prevented recovery of the vehicle. The SV-5D fell into the ocean at a high rate of speed.[50]

The second launch in the series, completed on March 5, 1967, performed a cross-range (lateral) maneuver of about 654 nautical miles that exceeded its planned performance. Air recovery was not attempted because a chute malfunction had placed the vehicle in a nose-down attitude rather than the nominal nose-up attitude. The backup water recovery system deployed as planned on impact, but the vehicle subsequently tore loose and sank before it could be recovered.[51]

The third PRIME vehicle was launched on April 19, 1967. After completing an extended cross-range maneuver of more than 710 nautical miles, a waiting JC-130B aircraft successfully snagged the vehicle at 12,000 feet, less than five nautical miles from its designated target area.[52] Despite the disappointing loss of the first two vehicles, the PRIME mission was so successful in yielding flight data that the Air Force cancelled the scheduled fourth flight.[53]

The PRIME mission demonstrated for the first time lifting-body technology, and succeeded in developing the world's first successful maneuverable reentry vehicle. It also evinced that systems developed for PRIME were capable of precision accuracy in guiding the SV-5D to a preselected point. Data was also collected on aerothermodynamic and heat shield technology at hypersonic speeds. Many of PRIME's accomplishments would find practical application during Space Shuttle development. The PRIME spacecraft is on exhibit at the Air Force Museum, Wright-Patterson AFB, Dayton, Ohio.[54]

## *A Snapshot of Launch Vehicle and Facility Accidents*

Not all of the launches were successful. Sometimes they exploded and burned on the launch pad or during powered flight. At other times missile flight safety officers would destroy the vehicle while in flight when it failed to remain on course or exhibited abnormal movements. Accidents involving personnel during missile training exercises proved to be the most lethal at Vandenberg AFB.

Much of the mayhem occurred during the early years of launch operations and has since become an infrequent aberration. Several factors have contributed to the improved safety record, including more reliable launch systems and modern instrumentation that detects issues during vehicle processing before they become irrevocable in-flight problems. Stricter safety requirements and years of accumulated launch experience have also enhanced the safety record. Finally, the current launch rate is a tiny a fraction of the booming 1960s rate, so the odds of experiencing a missile failure are statistically reduced. These measurements by themselves do not eliminate the inherent risk each time one of these multi-ton monsters loaded with volatile propellant is shot into the air. There is also no guarantee against disasters caused by human error, as witnessed in the Space Shuttle program.

**An Atlas D or Titan I ICBM headed for destruction shortly after liftoff from Vandenberg AFB. U.S. Air Force.**

The first missile loss at Vandenberg AFB from a launch attempt occurred during the sixth mission, on June 16, 1959. A Thor intermediate-range ballistic missile lifted off from Complex 75-2, Pad 7 (75-2-7), but instead of turning westward out over the Pacific Ocean, it ascended straight up to an altitude of at least 130,000 feet. At 106 seconds into the flight the missile flight safety officer pressed the destruct button, terminating the flight. The engine and other large parts of the missile fell into the ocean and were salvaged several weeks later by a U.S. Navy recovery team. Many smaller pieces of the airframe fell on the airbase or drifted up to twenty miles inland to land in Santa Maria and Guadalupe. One piece was reported to have struck the ground eight feet from a woman working in a field. The local press covered the incident in some detail, with photos of local residents holding pieces of the missile they found.[55]

On the night of July 22, 1960, Airman Second Class Richard Pierson was on duty at an Atlas D Complex (probably 576B-3) when he noticed excessive vapors escaping from the valve controlling the flow of liquid nitrogen in the tank storage pit. Intending to shut off the valve, he descended the ladder into the pit and shouted his location to SSgt. James Walker, who was working nearby. As the seconds passed, the pit was quickly fogging up with a deadly cloud of nitrogen vapor. Walker called out to Pierson. There was no reply. Then he heard the thud of a safety helmet striking the pit floor. Walker raced after Pierson and he too fell unconscious from the fumes that had replaced the breathable air in the pit. A third crew member, SSgt. Charles E. Thayer, had been observing the attempted rescue. Thayer called out to the men now concealed behind a curtain of dense vapor. Receiving no answer, Thayer dashed to the personnel shelter and informed acting pad chief MSgt. Carl F. Kurashewich. With TSgts. William R. Neely and Donald J. Sharpless, they ran to the scene. Kurashewich vaulted the side of the pit to avoid stumbling over the helpless crewmen. Holding his breath, and with critical seconds ticking away, he groped his way through eight feet of asphyxiating nitrogen and closed the valve. He then leaped from the pit and began spraying water over the base of the latter until the men were visible.[56]

Kurashewich directed his biggest man, Neely, to get Pierson first, since the junior airman had inhaled the heaviest does of nitrogen. The muscular Neely handed Pierson over the railing to Kurashewich and Sharpless. Kurashewich began administering artificial respiration to Pierson. By the time Walker was pulled to safety, Pierson had begun to breathe on his own. Kurashewich repeated the process for Walker.[57]

Walker and Pierson were taken by ambulance to the base hospital where they eventually recovered. Sergeants Thayer and Sharpless each received the Air Force Commendation Medal on March 27, 1961. For his selfless actions, Kurashewich received the Airman's Medal on September 8, 1961. Orders confirming awards for the other men could not be located.[58]

A little more than a year later a similar nitrogen accident occurred at Vandenberg that ended tragically in the death of Airman Second Class Robert L. Duncan on August 8, 1961. Around 10:00 in the morning, Duncan was part of a detail assigned to transfer liquid nitrogen from a tanker truck into a storage tank at Atlas E Complex 576C. Duncan was stationed inside the room where the nitrogen was being deposited. When the nitrogen overfilled the tank and began leaking from the tri-cock valve, which was mistakenly left open, Duncan called the crew member standing near the tanker to stop the nitrogen transfer. He then asked another airman to lock hands with him while he climbed part way down the storage pit ladder and leaned out and down to close the valve. By this time, much of the room was foggy and the pit blanketed by a thick layer of nitrogen gas. Suddenly, Duncan's arm went limp. Overcome by the gas, Duncan slipped away from the airman holding his wrist and fell to the bottom of the pit. A rescue party quickly donned respirators. By the time they pulled him from the pit the 25-year-old Duncan was dead from asphyxiation.[59]

Having experienced explosions and missile mishaps in its Thor and Atlas programs at Vandenberg AFB, the Air Force hoped for better luck with the Titan program. In November 1959 the Army Corps of Engineers turned over the first of four elevator-lift Titan I silos, the Operational System Test Facility

(OSTF), to the Air Force. Although mostly complete, basic construction continued until the end of May 1960, when Air Force contractors began installing and testing facility equipment. During this phase, numerous site deficiencies had postponed installation of an operational Titan in the silo until October 14. Subsequent delays during full system checkout forced the rescheduling of Vandenberg's first Titan launch from October to December.[60]

One of the first system checks of the subterranean Titan involved propellant loading. In mid–October 1960 a single load of liquid nitrogen was being pumped into the storage tanks when an equipment malfunction caused it to spill out onto the ground. At work in the equipment terminal were five members of a limited access crew who were unaware that the nitrogen was lowering the oxygen content in the air in their vicinity. At about 6:30 a.m., TSgt. Cecil L. Spears, a 1st Missile Division safety technician, and Mr. Mitchell O. Hoenig, the test conductor, donned respirators to check valve and relay positions in the area where the liquid nitrogen was being loaded into the tanks. About fifteen minutes later Spears telephoned the launch complex safety officer, Maj. James W. Schafer, about the nitrogen spill. He recommended an immediate evacuation of the area because the oxygen content in the access tunnel had fallen below the amount needed to support life.[61]

Schafer, who was in the silo control center at this time, put on a respirator and grabbed two additional 40-pound air breathing devices before rushing down the tunnel between the blast lock and the equipment terminal. In the terminal the extra devices were fitted on to two of the trapped technicians. Spears led the men through the 600-foot-long tunnel to safety. He then hurried back down the tunnel with the devices, and with the help of Guymond Kennedy, a Martin Company safety technician, they evacuated the remaining three crew members. Having given up his own respirator to safely evacuate the last of the employees, Schafer remained behind in the equipment terminal until a breathing device was brought back to get him through the tunnel. Just after Schafer was rescued, the oxygen level through much of the facility fell below human survival levels. Fortunately, no lasting ill effects were reportedly suffered by any of the men involved in the incident.[62]

The accident revealed numerous shortcomings in the facility and in safety procedures. Solutions to these problems included adding television cameras to allow the safety officer in the control center to read oxygen gauges in the propellant terminal, the installation of additional oxygen sensors, improved ventilation, and storage of emergency equipment at all hazardous locations. By the end of November, Air Force contractors had completed the safety upgrades. Ground testing of the missile system was also completed. The work now shifted to the final testing of airborne and ground equipment.[63]

On the night of December 3, 1960, a contractor/Air Force team fueled the Titan I with liquid oxygen and RP-1 fuel, then raised the missile on the silo-lift elevator to the surface. Some twenty-five operating personnel, including SAC observers, were in the underground OSTF control room at the time. At about 8:30 p.m. as the missile was being lowered back into the silo, the giant elevator failed. Both elevator and missile picked up momentum, and seconds later, after hitting bottom, the Titan propellant tanks ruptured. The spilling fuel ignited into a deflagration whose tremendous explosion sent base residents rushing out of their homes and rattled the nearby town of Lompoc.[64]

A shower of flame, steel, and concrete shot hundreds of feet into the air past the still open silo doors, which were severely damaged by the forces that were generated. The remains of Vandenberg's first Titan I operational missile, nicknamed "Gold Street," were scattered hundreds of feet from the silo. Personnel in the tunnel, on the safe side of the blast lock doors, and in the underground control center were shaken by the detonation but uninjured. Less than a month before the accident the existing blast doors were strengthened to withstand at least 300 pounds per square inch of overpressure. The decision to improve the doors probably saved the lives of the men in the complex. Above

Among the facilities under construction for the Titan program at Vandenberg AFB was the hardened underground silo-type Operational System Test Facility (OSTF). This November 16th sequence photograph illustrates the Titan I being elevated to the surface from its protective silo. The series of tests were virtually complete except for actual launch when, on December 3, 1960, the elevator malfunctioned while lowering the missile into the silo. The facility was destroyed by an explosion when the missile, fully loaded with propellant, and elevator fell to the bottom of the silo. U.S. Air Force.

ground, the missile accident emergency team went into action to seal off the area, and firefighting equipment was dispatched. Despite the catastrophic nature of the accident, miraculously there were no injuries.[65]

Air Force investigators probing the site in the days after the incident confirmed earlier reports that restoration was impractical. Much of the silo was later salvaged and then abandoned. Today, wildlife inhabit the crumbling silo ruins, which are scarcely visible through the dense brush that has grown up around the facility.[66]

On September 21, 1964, maintenance personnel from McConnell AFB and the 395th Strategic Missile Squadron at Vandenberg AFB were working on levels 3 and 5 of Titan II Complex 395D, conducting a final checklist inspection of the launch duct and work platforms mechanisms before placing missile B-32 on alert-ready status. Also in the group was Richard H. Rector, a civilian safety technician with the Navy's Pacific Missile Range, assigned to Vandenberg AFB. On duty in the launch control center (LCC) were Maj. R.

The aftermath of the Titan I explosion at the OSTF. The crib and elevator were blown out of the silo and the surrounding area pockmarked with debris. The Air Force declared the facility a complete loss. Fortunately there were no deaths or injuries to personnel at the site. U.S. Air Force.

O. Arnold, missile combat crew commander; Capt. L. L. Miller, deputy combat crew commander; SMSgt. G. E. Hensley, ballistic missile analyst technician; and TSgt. O. H. Weinhold, missile facilities technician. At about 8:13 p.m. Maj. Arnold directed TSgt. Weinhold to align the missile guidance system from heat to ready mode. Almost immediately after he pressed the button on the Missile Guidance Alignment and Checkout Group console, personnel on level two of the LCC felt a rush of air and heard blast door 9 slam shut.[67]

The missile's stage II translation rocket motor 2 ignited, sending a surge of air pressure rushing through the silo. The airmen on level 3 of the launch duct, the level of the translation rocket, were hit by a partially deflected blast and intense light from the rocket exhaust. SSgt. Adriano Ancheta suffered severe burns to his hands and fingers, and abrasions to the corneas of his eyes. SSgt. Charles B. Smith also sustained abrasions to the cornea, and second degree burns. A2C Edwin C. Bartz sustained second degree burns on the neck and ears. (Some Air Force reports place two other airmen on this level, SSgt. Gale L. McGregor, and SSgt. James M. Stowe.)[68]

On level 5, Rector was the most seriously injured. The force of the exiting air pressure hurled him through the open blast lock door and slammed him against the huge air conditioner. His injuries included a fractured tibia and a severe laceration on his left leg. A2C Charles R. Degenhart (or Dagenhart) was also tossed past the door. He sustained a lacerated right knee, various contusions and abrasions on the right leg, and a sprained right wrist. A2C James K. Wobser, and A2C Dennis L. Hockman incurred abrasions and contusions when the blast knocked them into various pieces of solid equipment. (McGregor, who experienced minor hair singing, and Stowe may have been on this level.)[69]

For approximately ten seconds until the translation rocket motor automatically shut down, the overpressure kept the doors tightly closed while smoke and flames spewing from the missile filled the cramped silo with the hot, acrid stench of propellant. Immediately after the motor had stopped firing, the men scrambled for the exits, assisting each other to the second level of the LCC where they received first aid before being evacuated to the hospital. Meanwhile, the evacuation klaxon sounded, the water deluge system started, and all emergency backout procedures were initiated. The accident could have been catastrophic had the flames reached the missile's ordnance or propellant tanks, or started a fire in the silo.[70]

An accident investigation board formed after the incident determined the primary cause of the motor firing was an electrical short in the igniter assembly of translation rocket motor 2. When the guidance system was advanced to the ready mode, an unexpected voltage surge caused the igniter to fire.[71]

Meanwhile, the stage II motor was removed from the missile to repair a

**A Titan II missile in a silo at Vandenberg AFB. The work platforms around the missile are in their upright position in preparation for launch. U.S. Air Force.**

hole burned in the missile skin and to replace damaged wiring. The stage was reinstalled on October 15, and the entire missile system underwent system checks.[72]

The refurbished missile was launched on November 4, 1964, and successfully sent a reentry vehicle ripping through the atmosphere to the Kwajalein

Missile Range in the Pacific Ocean. A Strategic Air Command missile combat crew from the 532nd Strategic Missile Squadron at McConnell AFB, Kansas, conducted the training mission.[73]

The "trailer court" incident became the most controversial missile accident at Vandenberg AFB. On September 2, 1965, at 1:00 p.m. the 6595th Aerospace Test Wing launched a Thor-Agena D rocket from Complex 75-3, Pad 5, carrying a secret spacecraft. The vehicle began drifting off course, crossed the abort line, and sixty seconds after liftoff was destroyed by the missile flight safety officer. Falling from an altitude of approximate 32,000 feet, the main parts of the rocket and burning propellant landed on the base and started multiple grass fires. Smaller pieces of debris drifted at least seven miles inland to the communities of Vandenberg Village and Mission Hills. Mrs. James P. Meachum and her four children narrowly escaped injury or death when an eight-foot-long section of the booster collapsed a portion of their house trailer

A section of the Thor-Agena D rocket launched from Vandenberg on September 2, 1965, landed on a house trailer on base, narrowly missing the trailer's occupants. The Air Force missile flight safety officer destroyed the rocket after it began drifting off course. Pieces of the missile were scattered over a wide area. U.S. Air Force.

at the Ocean View Trailer Park at Vandenberg AFB. Distraught by this harrowing experience, Mrs. Meachum was transported to the base hospital for observation and was released later in the day. A second trailer, owned by Airman James H. Parana, was also hit by falling debris but sustained only minor damage. It was unoccupied at the time.[74]

The accident investigation board stated in its report of September 24, 1965, that wind was the primary cause for the vehicle to cross the abort line, and that inaccurate pre-launch trajectory data was a contributing factor. Headquarters Air Force reversed this main finding and pinned the accident primarily on incomplete pre-launch trajectory data that failed to take into effect the full effects of wind on the launch vehicle. The Thor-Agena for this mission was configured with an oversized payload shroud 44 inches longer than the standard vehicle which made it more susceptible to wind effects. The accident and its findings resulted in a major overhaul of range safety launch procedures at Vandenberg.[75]

## The Cuban Missile Crisis and Vandenberg AFB

On October 22, 1962, Vandenberg AFB, along with the rest of America's defense establishment, went to an increased state of military readiness under Defense Condition 3 (DEFCON 3). The condition was prompted by what became known as the "Cuban Missile Crisis" after Soviet ballistic missiles were detected in Fidel Castro's island nation. Before the crisis ended on November 6, United States forces achieved nearly maximum readiness under DEFCON 2.[76]

Under the emergency DEFCON 3 status at Vandenberg AFB, command posts were set up and went into around-the-clock operations. As the situation over Cuba became more unstable, the Vandenberg team of Air Force and missile contractors began filling available launch facilities with ICBMs and preparing them with the appropriate combat hardware. In line with the increased readiness, all military leaves were cancelled, personnel on leave were recalled to duty, and all other personnel reassignments were temporarily suspended. Only personal emergency leaves were authorized. Aside from cancelling leave, outwardly and in the base newspaper, Vandenberg AFB appeared calm and continued its normal daily activities. On October 24, 1962, two days before launching a Thor-Agena D space booster and an Atlas D ICBM, the Air Force issued a press release reassuring the public the launches had been scheduled weeks earlier and were unrelated to the unfolding situation in Cuba. Not until much later would the American public learn how perilously close the United States and the Soviet Union had come to nuclear annihilation.[77]

## *Guess Who's Coming to Vandenberg*

Since its establishment in 1957, Vandenberg AFB has greeted thousands of visitors, some famous, some distinguished, some both, and many more neither of the two. Probably the first foreign head of state to pass through Vandenberg AFB—though technically an unintended guest of the U.S. Government—was the portly and bombastic Russian premier Nikita S. Khrushchev. His trip came about in July 1959 when U.S. Undersecretary of State Robert Murphy misinterpreted instructions he received from President Eisenhower and passed these on to his Russian counterpart, where it was received as an invitation to visit the United States. Khrushchev gladly accepted the "invitation." Eisenhower had indicated to Murphy that he was willing to meet with the Premier at Camp David in Maryland if progress was made in talks between the United States, Britain, and the Soviet Union over a crisis the Soviet Union manufactured a year earlier concerning Berlin. Eisenhower was furious over Murphy's foul-up, but nothing diplomatically could be done to reverse the situation. He had to accept that Khrushchev would be traveling across the United States for two weeks.[78]

On September 15, 1959, Chairman Khrushchev and his entourage flew into Washington D.C. in what was billed as a good will tour. After meeting with Eisenhower at the White House and attending some perfunctory meetings with other officials, Khrushchev went by train over a heavily guarded route to New York for additional sightseeing and a speech at the United Nations. He then flew to Los Angeles where he received a frosty reception from Mayor Norris Poulson. Khrushchev was then taken by motorcade to the Beverly Hills studio of Twentieth Century–Fox for lunch. At the banquet, and surrounded by Hollywood's leading actors, he got into a verbal debate with the studio president, Spyros P. Skouras, involving the virtues of capitalism over communism. When he received word that American security and local police officials had reneged on a promised visit to Disneyland because they could not guarantee his safety, Khrushchev expressed his displeasure.[79]

After lunch, Khrushchev was taken to Sound Stage 8 to watch the filming of *Can Can*, which he apparently enjoyed but later denounced as capitalist decadence. Later that evening at a cocktail party at the Ambassador Hotel ballroom, Mayor Poulson delivered a speech that enraged Khrushchev, who responded in kind.[80]

The following morning, on September 20, Khrushchev boarded a Southern Pacific locomotive to San Francisco, with whistle stops in Santa Barbara and San Luis Obispo, where he was greeted by friendly crowds. Traveling with Khrushchev was Henry Cabot Lodge, American Ambassador to the United

Nations, who was selected by Eisenhower to be Khrushchev's tour guide across America.[81]

Knowing the train would pass through Vandenberg AFB, the U.S. State Department wanted Khrushchev to see America's first generation of ICBMs standing on alert. As the train rolled through Vandenberg along the Pacific coast, Khrushchev left his seat, strolled to the press car, and sat with his back to the window. When the two shiny silver Atlas missiles at Complex 576A came into sight and were announced over the loudspeaker system, a reporter informed Khrushchev. Without even glancing over his shoulder, he retorted, "I'm not interested in your missiles, we have more than you do and ours are better."[82]

Khrushchev almost certainly knew that had he turned to look at the missiles, the press would have had a field day describing his facial expression. And in the realm of diplomatic politics, he then would have been expected to reciprocate with a visit to a Russian missile base for President Eisenhower, something the Soviet Union would never tolerate.[83]

The vice president of the United States, Lyndon B. Johnson, visited the Naval Missile Facility Point Arguello and Vandenberg AFB on October 4, 1961. He received an orientation on their respective missions and tours of the two installations. Johnson's California agenda included stops at the Ames Research Center, the Jet Propulsion Laboratory, and Edwards AFB. As Chairman of the National Aeronautics and Space Council, this was the first of Johnson's "listen and learn" trips, as he called them.[84]

A special locomotive carrying Russian premier Nikita Khrushchev passes Atlas Launch Complex 576A and the Strategic Air Command's sign, 1st Missile Division, on its way through Vandenberg AFB. Khrushchev turned his back to the missiles visible from the train. September 20, 1959. U.S. Air Force.

Five months later, on March 23, 1962, the attorney general of the United States, Robert F. Kennedy, flew into Vandenberg AFB aboard a Navy aircraft. Kennedy made a brief inspection of the missile display set up on the side of the runway for his brother's tour later that afternoon. He then went directly to the Federal Correctional Institution, the former U.S. Army Branch Disciplinary Barracks, near Lompoc and conferred with Warden Greig V. Richardson. The attorney general visited the workshops, housing quarters, recreation grounds, chapel, and other areas of the prison. At the end of his two-hour tour, Kennedy departed for Los Angeles to attend on the next day the Conference on Crime Prevention in California.[85]

President John F. Kennedy and his entourage arrived at Vandenberg AFB by jet at 4:08 p.m. on March 23, 1962. As the President stepped from the plane and paused for a moment, spectators behind a barrier of military police cheered. A polished Air Force honor guard snapped to attention, and the Fif-

President Kennedy watches Atlas 134D ascend into the skies over Vandenberg AFB on March 23, 1962. Directly behind the President is Harold Brown. Also in the back row is Roberts S. McNamara and Maj. Gen. Joseph J. Preston. In front, beside the President, is Gen. Thomas S. Power. U.S. Air Force.

**Atlas 134D lifted off from Complex 576 B-2 and performed flawlessly. March 23, 1962. U.S. Air Force.**

teenth Air Force band from March AFB greeted the President with a rousing rendition of "Hail to the Chief" and the "National Anthem." With Kennedy were Secretary of Defense Roberts S. McNamara; director of defense, research and engineering at the Department of Defense, Harold Brown; the commander in chief of the Strategic Air Command, Gen. Thomas S. Power; and the President's press secretary, Pierre Salinger, with some fifty members of the White House Press Corps. Maj. Gen. Joseph J. Preston, commander of the 1st Strategic Aerospace Division, greeted them at Vandenberg.[86]

The President arrived from Berkeley, California, where earlier in the day he received an honorary degree from the University of California and gave the Charter Day address commemorating the anniversary of its founding.[87]

At Vandenberg, Gen. Power guided the presidential party on a tour of the base's missile facilities. The highlight of the visit came about thirty minutes after the President arrived when he witnessed the launch of an Atlas missile. The tour got underway with Kennedy examining an Atlas D, Thor-Agena, Titan I, and Minuteman I display on the side of the runway. He then entered the black Lincoln limousine with its clear bubble top that took him to the 576A missile complex. From this observation point overlooking a valley, President Kennedy watched and smiled with exuberance as Atlas 134D lifted off from Complex 576B-2 and soared into the clear blue skies over Vandenberg AFB. A Strategic Air Command missile combat crew from the 389th Strategic Missile Wing, Warren AFB, Wyoming, launched the vehicle, nicknamed "Curry Comb I," at 4:39 p.m. The missile traveled down the Pacific Missile Range and successfully delivered a Mark 3 reentry vehicle 4,386 nautical miles to a target in the Eniwetok Lagoon.[88]

All three pads at Complex 576B were fitted with missiles for the presidential visit. Raised from horizontal storage to a vertical firing position and placed in a ready state, they projected a spectacular tableau of strength and national pride. The terminal countdown for Atlas 134D was halted after the missile was fully loaded with propellant to allow the final two minutes of the countdown to be timed with the arrival of the presidential party at the viewing site. The additional two missiles at 576B served as available backups if the first missile had failed. They could have been launched in approximately one-hour intervals, the time needed to reprogram instrumentation.[89]

After the launch the President's tour continued past gleaming Atlas F missiles at Complex 576D to a Minuteman launch control center where the President descended into the underground facility. While in the Minuteman area, Kennedy also inspected a Minuteman silo and the transport-erector that was used to place a Minuteman missile in its silo. The next stop on the tour was a Titan I launch control center. After a short briefing about the weapon system and the control center, a Vandenberg Titan crew conducted a simulated countdown, and two Titan missiles were elevated to the surface from their underground silos so the President could better view them.[90]

The presidential convoy drove to the control center at Complex 576B where Atlas 134D had been launched. Kennedy congratulated the crew and spoke briefly with each member. Its commander, Maj. Clifford W. Simonson, presented a missile badge to the President and escorted him into the control center for a briefing about the launch and to discuss the instrumentation.[91]

Returning to the flight line, Kennedy viewed aerospace exhibits set up in the aircraft hangar. Among these artifacts were missile models and missile components, including a reentry vehicle and a Discoverer XIV data capsule. The President returned to his aircraft, which took off at 6:53 p.m. for Palm Springs, California, where he spent the weekend at Bing Crosby's home. Code-named "Project Sky Rocket" by the Air Force, this was the only presidential visit to Vandenberg AFB.[92]

# EPILOGUE

Life in the small, agricultural-based communities near Camp Cooke was never going to be the same after the establishment of the military camp in 1941. Like many other wartime facilities, Cooke brought prosperity to the area, providing good paying defense jobs and opening the workforce to large numbers of women. The camp also drew new employees to the area, which grew the local population. To illustrate this point, Lompoc's population of 3,379 in April 1940 jumped to 5,844 by August 1944. This latter figure was almost certainly higher during camp construction when many migrant workers arrived in the area. While no census numbers are available from the Lompoc government offices for the period of the Korean War, the next set of figures, for October 1957, indicate 6,665 residents. By December 1959 the population had surged to 13,914, due almost exclusively to the rapidly expanding workforce at Vandenberg AFB. Over the next several decades the town continued to expand, though increasingly with less influence from the military base and more from the development of new businesses. As of 2013, more than 43,000 residents claimed Lompoc as their home.[1]

On October 18, 1991, fifty years after Camp Cooke was activated, some two-hundred people attended a ceremony at Vandenberg AFB that included the dedication of a memorial site and an evening banquet to honor Army veterans who trained at Cooke during World War II. Among those in attendance were about one-hundred former soldiers and their spouses.[2]

The memorial consists of an M-47 Patton tank placed on a raised site at California and Nebraska Avenues. Erected in front of the tank is a large concrete, triangular-shaped monument, the symbol of the armored command, with plaques from the 5th, 6th, and 11th Armored Divisions embedded on top. The M-47 came into army service at the end of the Korean War.[3]

The 5th Armored Division Association had contacted Air Force leadership at Vandenberg in May 1990 to establish the memorial using a World War

II U.S. Army tank. The Air Force embraced the idea, and a month later both groups began planning for the event. The association recruited the assistance of the 6th and 11th Armored Division Associations. Unable to locate a World War II tank for display, they obtained the M-47 from an Army arsenal in Alabama and had it shipped by rail to Vandenberg AFB. They also provided the brass plaques. The Air Force selected and prepared the site, which included fabricating the concrete monument for the plaques.[4]

On September 1, 2000, fifty years to the day that the 40th Infantry Division was activated and ordered to Camp Cooke, more than 1,000 guests gathered at Vandenberg AFB to dedicate a memorial to the thousands of men who served in that unit from 1950 to 1953. The memorial consists of three parts made of South Dakota granite, weighing 25,000 pounds. The large, tall piece bears the infantry division insignia and has a short history of the unit inscribed upon it. This represents the unit at Camp Cooke. The second vertical piece, with a small concrete pagoda on top, memorializes the division in Korea. Connecting the two pieces is a bridge, symbolizing training in Japan. On the ground around the memorial are bricks etched with the names of division members. A small, split-face brick wall encircles the entire site. The memorial was privately funded and gifted to Vandenberg AFB. It is located next to the World War II memorial.[5]

Apart from the two Army memorials at Vandenberg AFB, almost all of the original Camp Cooke structures are gone, except for a few buildings that have been significantly remodeled. The railroad spur built in 1942 and used to bring tons of supplies into the camp and to transport thousands of service members was ordered removed by the base commander in 2006. Unable to find a salvaging company willing to buy the steel, the Air Force handed the project over to the California Army National Guard and U.S. Border Patrol, who planned to recycle the rails as vehicle barriers near El Centro, California, to help secure the border with Mexico. The first of two teams of Army Guardsmen began removing rail in June 2006. Iron Horse Preservation Society of Nevada, under contract to the California Guard, completed the work in February 2008. The company not only removed the remaining 720.4 tons of rail, but all of the wood ties. The company donated to the federal prison near Lompoc about 1,000 damaged or otherwise unsalable ties.[6]

Through the end of 2013, Vandenberg AFB launched 1,929 missiles and rockets. More than a third of these vehicles, 653 to be exact, occurred before December 31, 1966. In that year a record 123 vehicles lifted off from Vandenberg, more than at any time before or since. After 1966 the launch rate steadily declined to eight launches in 2004, and today struggles to break the double digits. Several factors have contributed to the decline, including fewer launch failures and extended satellite life that requires less demand for replacement vehicles, a research and development flight test program that is virtually extinct, and the overall rising cost of launch operations that has tended to discourage new programs.[7]

By 1965–66, America's first-generation ICBM force of Atlas, Titan I, and Minuteman I vehicles were no longer in active service. Each of these vehicles had completed their flight test programs at Vandenberg AFB and were deployed at operational bases. They were succeeded by the Titan II and Minuteman II, which in turn yielded to the more advanced Minuteman III and Peacekeeper missiles. Today, only the Minuteman III remains in service. Modified Thor and Atlas missiles used as space boosters were slower to be phased out, but the Air Force's transition to new, larger space-lift vehicles began with the Titan IIIB in 1966. The 1990s brought the still larger Titan IV that operated until 2005. The commercially operated Delta IV, Atlas V, and Falcon space launch vehicles are the main heavy-lift boosters currently used to carry national defense satellites into orbit.[8]

Compared with the early years at Vandenberg when Air Force crews conducted launch operations with occasional assistance from missile contractors, all of today's space launches from Vandenberg are performed by commercial contractor teams, leaving Air Force and government employees with no involvement in vehicle or satellite processing, and minimal support in the actual launch operation. The Minuteman III ICBM is the only vehicle at Vandenberg that is launched entirely by military crews. As part of the Force Development Evaluation program, operational Minuteman III missiles are periodically removed from one of three ICBM missile bases at F. E. Warren AFB, Wyoming; Malmstrom AFB, Montana; and Minot AFB, North Dakota, and shipped to Vandenberg for launch. At Vandenberg, these missiles are thoroughly processed and fitted with a command destruct system, and an inert reentry vehicle. The launch crew usually comes from the same base as the missile. Occasionally, instead of the ground launch crew issuing the missile launch commands, a crew aboard a specially equipped airborne launch control system aircraft delivers them. The majority of the Minuteman III flights are targeted to a broad ocean area near the Ronald Reagan Test Site at the Kwajalein Atoll, some 4,200 nautical miles southwest of California. The launches are conducted in a near operational environment, thus enabling the Air Force to continually verify missile reliability and accuracy, and to maintain launch crew proficiencies.[9]

# APPENDIX A:
# BRIG. GEN. PHILIP
# ST. GEORGE COOKE,
# 1809–1895

Philip St. George Cooke was born in Leesburg, Virginia, on June 13, 1809. At the age of fourteen, Cooke entered the United States Military Academy at West Point, New York, and graduated four years later, on July 1, 1827, as a 2nd lieutenant. He was assigned to the 6th Infantry Regiment at Jefferson Barracks in Missouri.[1]

His first combat experience came two years later in a series of skirmishes against Comanche and Kiowa Indians along the upper Arkansas River. At the conclusion of the expedition, Cooke and his military unit were sent to Cantonment Leavenworth (later Fort Leavenworth), Kansas. Here he met Rachel Wilt Hertzog, the daughter of a Philadelphia merchant, and married her less than a year later, on October 28, 1830.[2]

In 1832, Cooke volunteered for service in the Black Hawk War against Sac and Fox Indian tribes. In August of that year he fought in the Battle of Bad Ax River in Michigan Territory (later Wisconsin). In March 1833 he was promoted to 1st Lieutenant and assigned to the new 1st United States Dragoons Regiment (mounted infantry). After a foray into the unorganized Indian Territory, Cooke contracted malaria and possibly dysentery. He was sent to Carlisle Barracks, Pennsylvania, to recover and assigned to recruiting duty. In 1835 he was promoted to captain, and during this time received his license to practice law in Virginia. Except for an occasional assignment escorting merchants though Indian Territory, Cooke spent most of the next seven years performing garrison work at outposts in the Kansas and Oklahoma territories.[3]

In the summer of 1842, Cooke was back on the frontier assisting in the forced movement of Seminole Indians in the Oklahoma Territory. A year later, while escorting a caravan of merchants headed from Fort Leavenworth to Santa Fe, New Mexico, he disarmed a band of marauding Texans that were planning to raid the wagon train. Cooke's next assignment was a three-month expedition from Fort Leavenworth to Wyoming and back, escorting settlers and shoring up friendly relations with Indian tribes along the Oregon Trail.[4]

When the United States declared war on Mexico in May 1846, Cooke was under the command of Colonel Stephen Kearny at Fort Leavenworth. Kearny received orders to capture Santa Fe and then proceed into California. Cooke rode ahead of the main force and, at the direction of Col. Kearny, helped accomplish the surrender of the city without firing a single shot, in August 1846. In September, shortly after leaving the city for San Diego, California, Cooke was dispatched to Santa Fe and put in charge of leading a battalion of about 500 Mormon volunteers for service in California. The battalion had come from Nauvoo, Illinois, with military escort. Aware of the hardships that lay ahead, Cooke weeded out sick men and almost all the women and children, and sent them to Pueblo, Colorado, for the winter. On October 19 he led a reorganized battalion on its remaining 1,100-mile march through a hazardous wilderness, pioneering a new wagon road that others were to use in later years. They arrived in San Diego on January 29, 1847, and less than a month later Cooke was promoted to major in the 2nd United States Dragoons. In July the Mormon Battalion was discharged from active military service.[5]

**Brig. Gen. Philip St. George Cooke. Courtesy U.S. Army Military History Institute.**

In the summer of 1847, after a brief reunion with his family in St. Louis, Missouri, Cooke was ordered to Washington, D.C., where he was the chief witness at the court-martial trial that convicted Frémont of failing to obey orders in California. Cooke rejoined his regiment in Mexico City in April 1848, but by this time the war had ended. From October of that year until October 1852 he served as post commander and superintendent of cavalry recruiting at Carlisle Barracks, Pennsylvania. He was promoted to lieutenant colonel in March 1849 in recognition of his service in California.[6]

Cooke yearned for the adventures of the open frontier. In November 1852 he welcomed a transfer to Texas where he engaged the Lipan Indians in a brief skirmish. The following year Cooke was order to the New Mexico Territory and received permanent promotion to the rank of lieutenant colonel. The defeat of Mexico had left a power vacuum in the territory that was quickly filled by bands of Apaches attacking white settlers. Cooke arrived in the territory at the height of hostilities and for the next two years was involved in several bloody clashes with Jicarilla Apaches, Utes, and Mescaleros before they surrendered to reservations.[7]

In 1855, Cooke was sent to the Nebraska Territory where he helped subdue the Brulé Sioux at Blue River Creek. Assigned as commander of Fort Riley in the Kansas Territory between 1855 and 1856, he helped restore order between pro-slavery and free-soil factions during the period known as "Bleeding Kansas."[8]

In September 1857, as part of an expedition against Mormons in Utah, Cooke was ordered to move troops and supplies from Fort Leavenworth to Fort Bridger, northeast of Salt Lake City. The brutal 1,000-mile march was accomplished through severe winter weather that resulted in sickness and frostbite among many of the men, and the loss of many animals. Cooke was promoted to full colonel in June 1858, and in September of that year was granted sixty days of leave to Washington, D.C. While on leave he began writing a cavalry manual for American horse soldiers. As part of his research, he traveled to Europe in 1859 to observe Napoléon III's Italian campaign, which had concluded by the time he arrived. He completed *Cavalry Tactics* in 1860, and a year later it was adopted by the War Department.[9]

In the summer of 1860, Cooke was assigned commander of Fort Floyd (renamed Fort Crittenden) in the Utah Territory. Within a few months, the Civil War fractured the nation and divided his family and friends along regional loyalties. Cooke had three daughters and one son, John Rogers Cooke, who resigned his commission in the Union Army and became a Confederate cavalry general. His eldest daughter, Flora, married Confederate cavalry commander J. E. B. (Jeb) Stuart. The second daughter was married to an Army surgeon who also gave up his commission to serve the Confederacy. Only his youngest daughter and her husband remained by his side, loyal to the Union cause. Cooke's former friend and fellow officer in the early Dragoons, Jefferson Davis, headed the succession.[10]

In October 1861, Cooke was ordered to Washington D.C., and a month later was appointed a brigadier general in the Regular Army and given command of a cavalry brigade guarding the city. The brigade later became part of the Army of the Potomac and served in General George B. McClellan's Peninsula Campaign in 1862. During a daring raid behind Union lines by J. E. B. Stuart, Cooke was ordered by his commander to maintain his position despite his desire and readiness to pursue Stuart. Stuart's incursion was quickly followed by three battles collectively called the Seven Days Battles. At Gaines's Mill, near the Confederate capital of Richmond, Cooke led a controversial cavalry charge that further tarnished his reputation, and in July 1862 he quit the Army of the Potomac.[11]

The remainder of Cooke's wartime service was administrative. In rapid succession he served on retirement boards in Washington, D.C., and court-martial boards in St. Louis. In October 1863, Cooke commanded the occupied District of Baton Rouge, Louisiana. From May 1864 until March 1866 he was posted to New York as superintendent of the Army's recruiting service. In March 1865 he was promoted to major general.[12]

When his assignment ended in New York, Cooke returned to the western frontier to command the Department of the Platte, where he confronted hostile Sioux Indians contesting the westward expansion of settlers along the Bozeman Trail. Cooke's mishandling of the facts involving the massacre of Captain William J. Fetterman and some eighty men in December 1866 led to his reassignment back east to serve on promotion and retirement boards. In May 1869 he commanded the Department of the Cumberland, where he was engaged in the reconstruction of West Virginia, Kentucky, and Tennessee. In May 1870 he received his last appointment—to the Department of the Lakes, with headquarters in Detroit.[13]

On October 29, 1873, Gen. Philip St. George Cooke retired from the Army after 46 years of active duty, and a little more than half a century after he had entered West Point. In retirement, family matters occupied a great deal of the old general's time. Cooke and his son John Rogers finally reconciled in 1887. Less than four years later John Rogers died. Gen. Cooke closed his eyes for the last time in his home in Detroit, Michigan, on March 20, 1895.[14]

# APPENDIX B:
## CAMP COOKE AND THE
## RANCHO DE JESUS MARIA

In the early nineteenth century, when California was part of Mexico, local government officials apportioned large areas of land to Mexican settlers. One of these grants was the Rancho de Jesus Maria. More than a century later, Camp Cooke was built on portions of several Mexican land grants, including the Rancho de Jesus Maria. At more than 40,000 acres, Jesus Maria was the largest of the grants, and virtually all of it was confined within the camp.[1]

Rancho de Jesus Maria was originally issued to Lucas Olivera in April 1837 for his service in the Mexican army. He built an adobe ranch house on the property near the San Antonio Creek. A modest structure, it consisted of a kitchen and family room with a large fireplace in one corner, and a bedroom.[2]

Meanwhile, in 1831, Lewis T. Burton, a Kentucky fur trapper, arrived in California. He moved to Santa Barbara where he married into a prominent Mexican family, changed his first name to Luis, and gradually built a thriving business as a retail merchant. After California became part of the United States, he was elected to several local political offices.[3]

During the 1850s, Burton began to purchase land outside Santa Barbara, and in 1853 he acquired Jesus Maria. Apparently there were irregularities in this transaction, and Burton's legal title to the ranch was not confirmed until 1871 after prolonged litigation in United States courts. Burton and his son Ben held the ranch for almost half a century, first raising cattle and horses, and later adding large flocks of sheep. They continued to live in Santa Barbara and stayed at the former Olivera adobe house during their visits to the ranch. Much of the ranch was located on a mesa that became known as Burton Mesa.[4]

Throughout this time, American settlers continued to move into California and homestead areas along the Central Coast. In 1874 the town of Lompoc was founded. Soon after, the Burtons allowed a group of investors to construct a wharf near Purisima Point called Lompoc Landing. The wharf became the lifeblood for commerce in the area until the railroad arrived some twenty years later and improved shipping. In 1883, Ben Burton began selling off Jesus Maria. One-hundred and twenty acres were sold to the Southern Pacific Railroad, which pushed its coastal line across the ranch in 1896. Ben's wife sold the remaining acres in 1903.[5]

Since oil had already been discovered in the community of Casmalia a few miles north of the ranch, the Union Oil Company quickly snapped up the remaining Jesus Maria, believing that oil deposits lay hidden beneath its acres. In 1906 the company sold off the surface occupancy rights to the Jesus Maria Corporation while retaining the mineral rights to the land. Heading the corporation was Edwin Jessop Marshall.[6]

E. J. Marshall was a remarkable man who started at the bottom in railroading, worked his way to success, and branched out into ranching, banking, and oil development. His greatest interest was ranching, and he developed a virtual empire in California, Arizona, and Mexico. His largest holding, the Palomas Ranch in Mexico, covered two million acres.[7]

Marshall tied the Jesus Maria into his empire by means of the Southern Pacific Railroad. He shipped calves from the Palomas to the Jesus Maria for a few

**Edwin J. Marshall. U.S. Air Force.**

months of fattening before transporting them to Los Angeles for sale. Marshall also maintained a permanent cattle herd on the ranch, and leased portions of the land to tenant farmers. These leases required the tenants to grow feed for Marshall's cattle on a portion of their fields.[8]

Marshall added more buildings to the ranch headquarters, planted a thick eucalyptus windbreak around it, started windbreaks at several other places on the property, and added 7,600 acres of land in the Point Sal area to the ranch. He built several small reservoirs to improve the irrigation system. As the years passed, Marshall enlarged the old adobe ranch house slowly and haphazardly. He had a living room, three bedrooms, and a kitchen built onto the adobe structure. These additions matched neither the adobe nor each other.[9]

In 1932, at the age of 72, Marshall retired from the directorship of several corporations and closed his offices in Los Angeles. Moving his family, his manager, and secretaries to the Jesus Maria, he turned his ambitions toward making the ranch a showplace. Carpenters and masons were hired to build a new stucco wing and to stucco over the older walls—adobe and clapboard alike. More windows were cut into the walls of the original adobe portion, and a covered porch was built across its front. The roof beams were strengthened and tile placed over the entire structure to create a Spanish motif. Stained glass windows, redwood paneling, damask wall covering, and a heavy door were brought from the Marshall family home in Los Angeles. These changes completely transformed the ranch house.[10]

By this time, the nationwide economic depression began taking its toll. Wild brush appeared on most of the ranch's fields, made idle by low farm prices. Undiscouraged, Marshall sought to supplement the Jesus Maria's income by starting a guest ranch. In 1934 a guesthouse was built near the ranch house, and by May 1935 the venture was in full operation. Marshall called the guest section of his Rancho de Jesus Maria "Marshallia" after a flower that had been named in honor of Marshall's eighteenth century uncle, Herbert, a noted American botanist.[11]

Marshallia became a popular retreat enjoyed by the glamorous personalities from Los Angeles and the film industry. Only the wealthy could afford to stay at the ranch, even at Depression economy rates. For $8 to $15 per day, or $50 to $90 a week, guests received cottage accommodations, all meals, horseback riding, and picnic trips.[12]

When E. J. Marshall died in a Los Angeles hospital on March 4, 1937, his widow moved to San Francisco, leasing out the ranch for cattle grazing. Morgan S. Tyler and his wife Marion negotiated a separate lease for the Marshallia guesthouse in August 1939 and operated a dude ranch. The ranch became increasingly popular among the wealthy as prosperity gradually returned to

America. The dude ranch also attracted movie stars such as Rod Cameron, Maria Montez, and Jeanette McDonald.[13]

On October 21, 1941, the Marshall estate sold the entire Rancho de Jesus Maria and its buildings for $750,000 to the United States Government, which made it the core of a new Army training installation named Camp Cooke. The Tylers were hired to stay on and maintain Marshallia as family quarters for the camp commanding officer and his staff. By this time, and probably earlier, the entire ranch property was known as Marshallia.[14]

The Army redecorated the Marshallia buildings and made some minor modifications. Significant changes occurred in 1942 when a fire that started in the fireplace of the ranch house caused extensive damage to the building. In making the repairs, the fireplace and two walls of the original Olivera adobe family room were removed. The house was also modernized with a central heating system and redecorated throughout.[15]

In July 1947, a year after Camp Cooke had closed for the first time, the Army leased Marshallia to the Tyler family, who again operated it as a private guest ranch. Horseback riding, fishing, and several miles of beach were available to guests. Fewer guests came to the ranch than before the war, however, forcing the Tylers to give up the business in March 1949.[16]

When Camp Cooke reopened in August 1950, the Marshallia buildings were repaired and slightly remodeled to again house senior military families, including the camp commander. After the Army closed the camp for a second time in March 1953, the Marshallia buildings housed key Army personnel operating the Branch U.S. Disciplinary Barracks. Flocks of sheep and cattle again roamed over the Burton Mesa.[17]

In 1957 when the Air Force took over the northern half of Camp Cooke, it also acquired Marshallia. The Air Force made several changes to the property, perhaps most notable the construction of a golf course and clubhouse, and replacing the guesthouse built by E. J. Marshall with a parking lot. The barn, bunkhouse, and other utility buildings were all removed over the next several years at various times. Today only the ranch house remains and bears the Marshallia name.[18]

# APPENDIX C:
# COMMANDERS AND
# COMMANDANTS

Camp Cooke was activated by General Orders No. 87, Headquarters, Ninth Corps Area, on October 5, 1941. When the camp closed for the first time on June 1, 1946, a small detachment of army personnel was left behind to watch after the military reservation. Six months later, on December 16, the Branch U.S. Disciplinary Barracks (USDB) opened on post property and received the added responsibility for all of Camp Cooke and the caretaker detachment. Shortly after the Korean War began, Camp Cooke was reactivated on August 7, 1950, and nine days later separated from the USDB. Col. Ben R. Jacobs, commandant of the USDB and commander of Camp Cooke during its caretaker phase, relinquished command of the camp to Col. Frank R. Williams. On March 31, 1953, Cooke closed for a second and final time. The camp was again turned over to the USDB under the command of Col. Benjamin B. Albert. All assigned dates listed below are while at Camp Cooke.

| *Camp Cooke Commanders* | *Dates of Assignment* |
| --- | --- |
| Lt. Col. John B. Madden | October 16, 1941–January 7, 1942 |
| Col. Carle H. Belt[1] | January 7, 1942–September 25, 1943 |
| Col. Harry C. Brumbaugh | September 25, 1943–May 1, 1945 |
| Col. Clayton J. Herman | May 1, 1945–May 1946? |
| Col. Hurley E. Fuller | May 9, 1946–August 1946 |
| Lt. Col. William S. Fowler | August 1946–November 16, 1946 |
| Capt. William (or Edward) T. Baldwin[2] | November 16, 1946–December (?) 1946 |
| Col. Frank R. Williams | August 16, 1950–November 19, 1951 |
| Col. Alexander G. Kirby | November 19, 1951–March 31, 1953 |

| *USDB Commandants* | *Dates of Assignment* |
| --- | --- |
| Maj. James M. McCarthy (Interim supervisor) | November 6, 1946–December 5, 1946 |

| USDB Commandants | Dates of Assignment |
|---|---|
| Col. James W. Fraser | December 5, 1946–May 19, 1947 |
| Col. Edward A. Everitt | May 19, 1947–March 15, 1950 |
| Col. Ben R. Jacobs | March 15, 1950–March 8, 1952 |
| Col. Lucian Berry | March 8, 1952–June 21, 1952 |
| Col. Harold Doud | June 21, 1952–July 1952 |
| Col. Benjamin B. Albert | July 1952–February 26? 1954 |
| Col. Harry L. Mayfield | February 26? 1954–May 25, 1955 |
| Col. Sam F. Muffie | May 25, 1955–December 20, 1957 |
| Col. Weldon W. Cox | December 20, 1957–August 1, 1959 |

| 5th Armored Division | Dates at Camp Cooke |
|---|---|
| Maj. Gen. Jack W. Heard | February 1942–March 1943 |
| Maj. Gen. Lunsford E. Oliver | March 1943–March 1943 |

*6th Armored Division*

| | |
|---|---|
| Maj. Gen. H. H. Morris, Jr. | March 1943–May 1943 |
| Maj. Gen. Robert W. Grow[3] | May 1943–January 1944 |

*11th Armored Division*

| | |
|---|---|
| Maj. Gen. Edward H. Brooks | February 1944–March 1944 |
| Maj. Gen. Charles S. Kilburn | March 1944–September 1944 |

*13th Armored Division*

| | |
|---|---|
| Brig. Gen. Wayland B. Augur | September 1945–November 1945 |
| Maj. Gen. John Millikin | July 1945–September 1945 |

*20th Armored Division*

| | |
|---|---|
| Maj. Gen. John W. Leonard | August 1945–March 1946 |

*86th Infantry Division*

| | |
|---|---|
| Maj. Gen. Harris M. Melasky | September 1944–November 1944 |

*97th Infantry Division*

| | |
|---|---|
| Brig. Gen. Milton B. Halsey | September 1944–February 1945 |

*40th Infantry Division*

| | |
|---|---|
| Maj. Gen. Daniel H. Hudelson | September 1950–March 1951 |

*44th Infantry Division*

| | |
|---|---|
| Maj. Gen. Harry L. Bolen | February 1952–December 1952 |

# APPENDIX D:
## KNOWN ARMY UNITS
## AT CAMP COOKE

    Every effort has been made to identify units assigned to Camp Cooke during World War II and the Korean War, and to align them under the correct chain of command. Unit reorganizations, redesignations, incomplete official records, and a large number of detached units during these conflicts have made this listing neither complete nor perfect. Tactical units of the Army Ground Forces, including National Guard components inducted into federal service, often changed in composition between the time they were in training camps and their overseas wartime service. For that reason, units at the division level mentioned in this history will show variations from lists compiled in the Order of Battle. Finally, this list does not include units that conducted summer training exercises at Cooke during periods of camp inactivation. Units that are known to have visited Cooke during those times are mentioned in the narrative. The following list of abbreviations is used in this appendix:

**AAA**—Antiaircraft Artillery
**AM**—Automotive Maintenance
**ASU**—Area Service Unit
**AW**—Automatic Weapons
**Bde**—Brigade
**Bn**—Battalion
**Btry**—Battery
**CA**—Coast Artillery
**CASC**—Corps Area Service Command
**CC**—Combat Command
**CID**—Criminal Investigation Detachment

**Co**—Company
**Comd**—Command
**Cons**—Construction
**Det**—Detachment
**Div**—Division
**Engr**—Engineer
**Gp**—Group
**HA**—Heavy Auto
**HHB**—Headquarters & Headquarters Battery
**HHC**—Headquarters & Headquarters Company
**HM**—Heavy Maintenance

**Hqtrs**—Headquarters
**Inf**—Infantry
**LM**—Light Maintenance
**Maint**—Maintenance
**MAM**—Medium Automotive
    Maintenance
**Mecz**—Mechanized
**MM**—Medium Maintenance
**MP**—Military Police
**MRU**—Medical Reserve Unit
**QM**—Quartermaster
**Rcn**—Reconnaissance
**Regt**—Regiment
**RR**—Railroad
**SCU**—Service Command Unit
**Sqdn**—Squadron
**TD**—Tank Destroyer
**Trk**—Truck
**Trp**—Troop
**WAC**—Women's Army Corps

## 1941

### Station Headquarters
CASC 1908
    (Redesignated SCU 1908 in July 1942)

## 1942

### 5th Armored Division
Headquarters
Hqtrs Co, 5th Armored Div
Hqtrs Co, Div Trains
Service Co
Supply Bn
Maintenance Bn
34th Armored Regt
81st Armored Reg
46th Armored Infantry Reg
22nd Armored Engineer Bn
58th Armored Field Artillery Bn
65th Armored Field Artillery Bn
75th Armored Medical Bn
85th Armored Reconnaissance Bn
145th Armored Signal Co
58th Ordnance Reg
58th Quartermaster Bn

### Miscellaneous Units
47th Quartermaster Co

705th Tank Destroyer Bn
807th Tank Destroyer Bn (Heavy SP)
815th Tank Destroyer Bn (Heavy SP)
816th TD Bn (Heavy SP)
199th Ordnance, Hqtrs and Hqtrs Det
86th Ordnance Co (HM, Tank)
23rd Quartermaster Reg, Co C, (Trk)
204th Quartermaster Bn, Co C, (4th
    Army)
223rd Quartermaster Bn
139th Quartermaster Co (Trk)
3rd Observation Squadron
166th AAA Gun Bn
168th AAA Gun Bn

## 1943

### 6th Armored Division
Headquarters
Hqtrs Co, 6th Armored Div
HHC Combat Command A
HHC Combat Command B
9th Armored Infantry Bn
44th Armored Infantry Bn
50th Armored Infantry Bn[1]
86th Cavalry Recn Sqdn (Mecz)
25th Armored Engineer Bn
146th Armored Signal Co
128th Armored Field Artillery Bn
212th Armored Field Artillery Bn
231st Armored Field Artillery Bn
128th Ordnance Maint Bn
76th Armored Medical Bn
15th Tank Bn
50th Tank Bn
68th Tank Bn
69th Tank Bn[2]

### Miscellaneous Units
2nd Filipino Infantry
2nd Hqtrs Special Troops, 4th Army
2nd Armored Inf Gp Hqtrs & Hqtrs
    Det
3rd Replacement Depot
9th Postal Regulating Section
11th Replacement Depot
13th Postal Regulating Station
14th Medical Depot Co
15th Tank Destroyer Group

XVIII Corps Artillery Hqtrs & Hqtrs
  Btry
31st Medical Group
33rd Field Hospital
34th Field Hospital
45th Replacement Bn
47th Replacement Bn
48th Replacement Bn
49th Replacement Bn
55th Signal Repair Co
62nd Signal Radio Interception Co
65th Replacement Bn
68th Replacement Bn
69th Medical Det
99th Quartermaster Bakery Bn
105th Evacuation Hospital
116th Antiaircraft Artillery Gp
116th Army Postal Unit
116th Coast Artillery Gp (AAA)
117th Army Postal Unit
128th General Hospital
130th Field Artillery Bn
137th Field Artillery Gp
144th Army Postal Unit
145th Army Postal Unit
154th Engineer Combat Bn
155th Engineer Combat Bn
156th Engineer Combat Bn
165th AAA Gun Bn
166th AAA Gun Bn
168th AAA Gun Bn
170th Field Artillery Bn
190th Ordnance Bn
193rd QM Gas Supply Co
198th Ordnance Bn
202nd AAA (AW) Bn
202nd Gas Supply Bn
207th Military Police Co
240th Station Hospital
258th Station Hospital
280th Field Artillery Bn
281st Field Artillery Bn
301st Balloon Barrage Bn
306th Quartermaster Bn
313th Balloon Barrage Bn
326th Ordnance Bn
341st Ordnance Depot Co
342nd Medical Group
382nd AAA (AW) Bn

412th Field Artillery Gp
  (Hqtrs & Hqtrs Btry)
441st Ordnance HA Maint Co
461st Ordnance Evacuation Co
483rd Ordnance Evacuation Co
488th Quartermaster Depot
496th AAA Gun Bn
506th AAA Gun Bn (Small)
529th Armored Infantry Bn
530th Armored Infantry Bn
531st Armored Infantry Bn
535th Armored Infantry Bn
540th Ordnance Co
553rd Army Postal Unit
554th Army Postal Unit
559th Ordnance HM Tank Co
561st Army Postal Unit
561st Ordnance HM Tank Co
566th Quartermaster Railhead Co
567th Quartermaster Railhead Co
568th Quartermaster Railhead Co
569th Quartermaster Railhead Co
579th Signal Depot Co
606th Tank Destroyer Bn
607th Clearing Co
607th Tank Destroyer Bn
618th Quartermaster Depot Co
705th Tank Destroyer Gp
708th Tank Bn (Amphibian)
716th Engineer Depot Co
773rd Tank Bn (Amphibian)
775th Tank Destroyer Bn
816th Tank Destroyer Bn
835th Quartermaster Gas Supply Co
836th Quartermaster Gas Supply Co
850th Ordnance Depot Co
879th Ordnance HA Maint Co
990th Engineer Treadway Bridge Co
1138th Engineer Combat Gp
1145th Engineer Combat Gp
1253rd Engineer Combat Bn
1254th Engineer Combat Bn
1468th Engineer Maint Co
1929th SCU:
  Branch School for Bakers and
  Cooks
1955th SCU:
  Ordnance Service Command Shop
3548th Ordnance HA Maint Co

3577th Quartermaster Trk Co
3578th Quartermaster Trk Co

## *1944*

### 11th Armored Division
Headquarters
Hqtrs Co, 11th Armored Div
HHC Combat Command A
HHC Combat Command B
HHC Division Trains
151st Armored Signal Co
56th Armored Engineer Bn
41st Cavalry Recn Sqdn (Mecz)
133rd Ordnance Maint Bn
81st Armored Medical Bn
11th Armored Division Band
Military Police Platoon
22nd Tank Bn
41st Tank Bn
42nd Tank Bn
490th Armored Field Artillery Bn
491st Armored Field Artillery Bn
492nd Armored Field Artillery Bn
21th Armored Infantry Bn
55th Armored Infantry Bn
63rd Armored Infantry Bn

### 86th Infantry Division
Headquarters
Hqtrs Co, 86th Infantry Div
341st Infantry Reg
342st Infantry Reg
343rd Infantry Reg
86th Div Artillery Hqtrs
331st Field Artillery Bn
332nd Field Artillery Bn
404th Field Artillery Bn
911th Field Artillery Bn
86th Cavalry Recn Trp
86th Quartermaster Co
311th Engineer Combat Bn
311th Medical Bn
786th Ordnance Co
86th Signal Co
86th Div Band
86th Military Police Platoon
86th Div Counter Intelligence Corps
  Det

43rd War Dog Platoon
  (attached to 86th Division)

### 97th Infantry Division
Headquarters
Hqtrs Co, 97th Infantry Div
97th Div Artillery Hqtrs
97th Division Band
97th Division Military Police Platoon
97th Division Quartermaster Co
97th Div Recn Trp (Mecz)
303rd Infantry Reg
386th Infantry Reg
387th Infantry Reg
303rd Field Artillery Bn
365th Field Artillery Bn
389th Field Artillery Bn
922nd Field Artillery Bn
332nd Engineer Combat Bn
322nd Medical Bn
797th Ordnance (LM) Co
45th War Dog Platoon
  (attached to the 97th Division)

### Miscellaneous Organizations
6th Italian QM Service Co
  (Italian Service Unit)
10th Tank Destroyer Group
22nd Replacement Depot
23rd Replacement Depot
24th OQ2A Radio Control Target
  Unit
25th AAA Gp, Hqtrs & Hqtrs Btry
28th AAA Gp, Hqtrs & Hqtrs Btry
33rd OQ2A Radio Control Target
  Unit
39th AAA (AW) Gp
46th Replacement Bn
47th Signal Heavy Construction Bn
51st Evacuation Hospital
51st Quartermaster Base Depot
54th Army Postal Unit
55th Army Postal Unit
56th Coast Artillery Reg (Mobile)
57th AAA Bde
  (Hqtrs &Hqtrs Btry)
66th Replacement Bn
74th Engineer Light Pontoon Co
75th Joint Assault Signal Co
  (Air Liaison Section)

77th Infantry Training Bn, Co A
88th Field Hospital
95th Replacement Bn
96th Replacement Bn
97th Replacement Bn
98th Replacement Bn
99th Evacuation Hospital
105th Evacuation Hospital
105th Army Postal Unit
106th Army Postal Unit
109th Replacement Bn
119th AAA Gun Bn
127th Ordnance Co
128th General Hospital
133rd Ordnance Maint Bn
142nd Italian QM Service Co
  (Italian Service Unit)
164th Station Hospital
167th Medical Bn
168th Medical Bn
169th Engineering Combat Bn
174th Ordnance
198th AAA (AW) Bn
234th Army Postal Unit
235th Army Postal Unit
259th Army Ground Forces Band
340th Ordnance Bn
381st AAA (AW) Bn
382nd AAA (AW) Bn
408th Field Artillery Gp
  (Hqtrs & Hqtrs Btry)
483rd AAA (AW) Bn
488th AAA (AW) Bn
505th AAA (AW) Bn
506th AAA Gun Bn
548th Ordnance HM Co
  (Field Artillery)
558th Ordnance HM Co (Tank)
563rd Quartermaster Railhead Co
631st MP Escort Guard Co
676th Engineer Light Equipment Co
710th Tank Bn
713th Tank Bn
722nd Engineer Depot Co
777th Infantry Cannon Co
778th AAA (AW) Bn
779th AAA (AW) Bn
782nd Tank Bn
786th AAA (AW) Bn

797th AAA (AW) Bn
798th AAA (AW) Bn
799th AAA (AW) Bn
815th AAA (AW) Bn
834th AAA (AW) Bn (SP)
871st Quartermaster Co
872nd Quartermaster Co
1134th Engineer Combat Gp
1416th Engineer Maint Bn
3435th Ordnance MAM Co
3436th Ordnance MAM Co
3522nd Ordnance MAM Co
3523rd Ordnance AM Co

## *1945*

### 13th Armored Division
Headquarters
Hqtrs Co, 13th Armored Div
HHC Combat Command A
HHC Combat Command B
Hqtrs Reserve Command
HHC Division Trains
83rd Armored Medical Bn
135th Armored Ordnance Maint Bn
Military Police Platoon
124th Armored Engineer Bn
153rd Armored Signal Co
24th Tank Bn
46th Tank Bn
16th Armored Infantry Bn
59th Armored Infantry Bn
67th Armored Infantry Bn
HHB Division Artillery
496th Armored Field Artillery Bn
497th Armored Field Artillery Bn
498th Armored Field Artillery Bn
93rd Cavalry Recn Sqdn (Mecz)
67th Order Battle Team
184th Photo Interpretation Team 513th
  Counter Intelligence Corps Det.

### 20th Armored Division
Headquarters
Hqtrs Company, 20th Armored Divi-
  sion
HHC, Combat Command A
HHC, Combat Command B
HHC Reserve Command

HHC, Division Trains
220th Armored Medical Bn
138th Armored Ordnance Maint Bn
Military Police Platoon
220th Armored Engineer Bn
160th Armored Signal Co
9th Tank Bn
20th Tank Bn
27th Tank Bn
8th Armored Infantry Bn
65th Armored Infantry Bn
70th Armored Infantry Bn
HHB Division Artillery
412th Armored Field Artillery Bn
413th Armored Field Artillery Bn
414th Armored Field Artillery Bn
33rd Cavalry Rcn Sqdn (Mecz)
74th Order Battle Team
191st Photo Interpretation Team
520th Counter Intelligence Corps Det.
46th Sound Locator Team
16th Recoilless Weapon Team

**Miscellaneous Organizations**
3rd Italian QM Co
  (Italian Service Unit)
XIII Corps, Hqtrs & Hqtrs Det
32nd Ordnance MM Co
37th Hqtrs & Hqtrs Det.
  (Special Troops, VII Corps)
Hqtrs & Hqtrs Det.
  (Special Troops, XXXVI Corps)
64th Army Ground Forces Band
88th Field Hospital
121st Ordnance MM Co
122nd Cavalry Rcn Trp (Mecz)
123nd Cavalry Rcn Trp (Mecz)
140th Cavalry Rcn Trp (Mecz)
140th Italian QM Co
(Italian Service Unit)
141st Cavalry Rcn Trp (Mecz)
170th QM Bn (Mobile)
172nd Engineer Combat Bn
178th Engineer Combat Bn
226th Signal Co
262nd Signal Light Cons Co
305th Signal Operations Bn
313th General Hospital
328th Ordnance Bn

339th Ordnance Bn
384th Ordnance Tank Maint Co
552nd Signal Aircraft Warning Bn
557th Field Artillery Bn
557th Ordnance HM Tank Co
561st Ordnance HM Tank Co
562nd Ordnance HM Tank Co
575th Ordnance AM Co
576th Ordnance AM Co
742nd Amphibian Tank Bn
764th Amphibian Tractor Bn
807th Tank Destroyer Bn
835th QM Gas Supply Co
839th QM Gas Supply Co
841st Ordnance Depot Co
987th Field Artillery Bn (SP)
990th Engineer Treadway Bridge Co
1017th Engineer Treadway Bridge Co
1021st Engineer Treadway Bridge Co
1169th Engineer Combat Gp
3351st QM Trk Co
3430th Ordnance MAM Co
3436th Ordnance Munitions Maint
  Co

### *August 1950–March 1953*

1st Signal Service Group (Sixth Army)
13th Armored Division
40th Infantry Division
  Headquarters
  Headquarters Co
40th Infantry Division Band
40th Military Police Co
40th Quartermaster Co
40th Replacement Co
40th Signal Co
40th Reconnaissance Co
40th Infantry Div Artillery
  (Hqtrs & Hqtrs Btry)
115th Medical Bn
140th Heavy Tank Bn
140th AAA (AW) Bn
143rd Field Artillery Bn
160th Infantry Regt
223rd Infantry Regt
224th Infantry Regt
578th Engineer Combat Bn
625th Field Artillery Bn

740th Ordnance Maint Co
980th Field Artillery Bn
981st Field Artillery Bn
44th Infantry Division
Headquarters
Headquarters Co
44th Signal Co
44th Military Police Co
44th Quartermaster Co
44th Division Band
44th Reconnaissance Co
44th Replacement Co
44th Division Artillery
123rd Field Artillery Bn
209th Field Artillery Bn
223rd Field Artillery Bn
233rd Field Artillery Bn
144th AAA (AW) Bn
106th Tank Bn
123rd Infantry Regt
129th Infantry Regt
130th Infantry Regt
135th Engineer Combat Bn
203rd Medical Bn
744th Ordnance Maint Co
47th Radio-Controlled Aerial Target Unit
49th Infantry Division
62nd Signal Radio Relay Co
91st Infantry Division
361st Infantry Regt
111th Armored Cavalry Regt
161st HM Ordnance Depot Co
303rd Signal Service Bn

306th Radio Broadcasting and Leaflet Gp
314th Signal Cons Bn
316th Military Police CID Det
317th Signal Cons Bn
325th Ordnance Gp
347th General Hospital MRU
349th General Hospital MRU
352nd General Hospital MRU
352nd Loudspeaker and Leaflet Co.
359th Engineer Battalion
362nd Engineers Gp
363nd Infantry Regt
375th Military Police Co
393rd Ordnance Battalion
394th Field Artillery Bn
421st AAA Gun Bn
450th Engineers Cons Bn
466th AAA (AW) Bn
499th Engineer Aviation Bde
747th Amphibious Tank & Tractor Bn
823rd Medical Station Hospital
838th Signal Radio Relay Co
881st Engineer Construction Bn
3265th Training Sqdn (Air Force) (Sixth Army Food Service Sub-School)
3623rd Ordnance Co (Medium Maint)
6014th ASU
6014th ASU U.S. Army Hospital
6014th ASU WAC Det.
6103rd ASU Disciplinary Barracks
6293rd Central Group, Det 21
3623rd Ordnance Co

# APPENDIX E:
# MEDAL OF HONOR RECIPIENTS
# WITH CONNECTIONS
# TO CAMP COOKE

Six soldiers assigned to units at Camp Cooke during World War II and the Korean War received the Medal of Honor for extraordinary acts of courage while in combat. A seventh man, Corporal Rodolfo P. Hernandez, did not train at Camp Cooke but was medically evacuated to the hospital at Cooke to recover from severe wounds received in Korea. The following information was obtained from the Congressional Medal of Honor Society Internet site at www.cmohs.org.

## *World War II*

**Burr, Herbert H.**
Rank and organization: Staff Sergeant, U.S. Army, Company C, 41st Tank Battalion, 11th Armored Division. Place and date: Near Dorrmoschel, Germany, 19 March 1945. Entered service at: Kansas City, MO. Birth: St. Joseph, MO. G.O. No.: 73, 30 August 1945.
**Citation**: He displayed conspicuous gallantry during action when the tank in which he was bow gunner was hit by an enemy rocket, which severely wounded the platoon sergeant and forced the remainder of the crew to abandon the vehicle. Deafened but otherwise unhurt, S/Sgt. Burr immediately climbed into the driver's seat and continued on the mission of entering the town to reconnoiter road conditions. As he rounded a turn he encountered an 88-mm antitank gun at pointblank range. Realizing that he had no crew, no one to man the tank's guns, he heroically chose to disregard his personal safety in a direct charge on the German weapon. At considerable speed he headed straight for the loaded gun, which was fully manned by enemy troops who had only to pull the lanyard to send a shell into his vehicle. So unexpected and daring was his assault that he was able to drive his tank completely over the gun, demolishing it and

causing its crew to flee in confusion. He then skillfully sideswiped a large truck, over-turned it, and, wheeling his lumbering vehicle, returned to his company. When medical personnel who had been summoned to treat the wounded sergeant could not locate him, the valiant soldier ran through a hail of sniper fire to direct them to his stricken comrade. The bold, fearless determination of S/Sgt. Burr, his skill and courageous devotion to duty, resulted in the completion of his mission in the face of seemingly impossible odds.

### Gammon, Archer T.

Rank and organization: Staff Sergeant, U.S. Army, Company A, 9th Armored Infantry Battalion, 6th Armored Division. Place and date: Near Bastogne, Belgium, 11 January 1945. Entered service at: Roanoke, VA. Born: 11 September 1918, Chatham, VA. G.O. No.: 18, 13 February 1946.

**Citation:** He charged 30 yards through hip-deep snow to knock out a machine gun and its 3-man crew with grenades, saving his platoon from being decimated and allow-ing it to continue its advance from an open field into some nearby woods. The platoon's advance through the woods had only begun when a machine gun, supported by rifle-men, opened fire and a Tiger Royal tank sent 88mm shells screaming at the unit from the left flank. S/Sgt. Gammon, disregarding all thoughts of personal safety, rushed forward, then cut to the left, crossing the width of the platoon's skirmish line in an attempt to get within grenade range of the tank and its protecting foot troops. Intense fire was concentrated on him by riflemen and the machine gun emplaced near the tank. He charged the automatic weapon, wiped out its crew of 4 with grenades, and, with supreme daring, advanced to within 25 yards of the armored vehicle, killing two hostile infantrymen with rifle fire as he moved forward. The tank had started to with-draw, backing a short distance, then firing, backing some more, and then stopping to blast out another round, when the man whose single-handed relentless attack had put the ponderous machine on the defensive was struck and instantly killed by a direct hit from the Tiger Royal's heavy gun. By his intrepidity and extreme devotion to the task of driving the enemy back no matter what the odds, S/Sgt. Gammon cleared the woods of German forces, for the tank continued to withdraw, leaving open the path for the gallant squad leader's platoon.

### Hastings, Joe R.

Rank and organization: Private First Class, U.S. Army, Company C, 386th Infantry, 97th Infantry Division. Place and date: Drabenderhohe, Germany, 12 April 1945. Entered service at: Magnolia, OH. Birth: Malvern, OH. G.O. No.: 101, 8 November 1945.

**Citation:** He fought gallantly during an attack against strong enemy forces defend-ing Drabenderhohe, Germany, from the dug-in positions on commanding ground. As squad leader of a light machine gun section supporting the advance of the 1st and 3d Platoons, he braved direct rifle, machine gun, 20mm, and mortar fire, some of which repeatedly missed him only by inches, and rushed forward over 350 yards of open, rolling fields to reach a position from which he could fire on the enemy troops. From this vantage point he killed the crews of a 20mm gun and a machine gun, drove several enemy riflemen from their positions, and so successfully shielded the 1st Platoon, that it had time to reorganize and remove its wounded to safety. Observing that the 3d Platoon to his right was being met by very heavy 40mm and machine gun fire, he ran 150 yards with his gun to the leading elements of that unit, where he killed the crew of the 40mm gun. As spearhead of the 3d Platoon's attack, he advanced, firing his gun held at hip height, disregarding the bullets that whipped past him, until the assault had carried 175 yards to the objective. In this charge he and the riflemen he led killed or wounded many of the fanatical enemy and put two machine guns out of action. Pfc. Hastings, by his intrepidity, outstanding leadership, and unrelenting determination to wipe out the formidable German opposition, cleared the path for his company's advance into Drabenderhohe. He was killed four days later while again supporting the 3d Platoon.

## Korean War

### Bleak, David B.

Rank and organization: Sergeant, U.S. Army, Medical Company 223d Infantry Regiment, 40th Infantry Division. Place and date: Vicinity of Minari-gol, Korea, 14 June 1952. Entered service at: Shelley, ID. Born: 27 February 1932, Idaho Falls, ID. G.O. No.: 83, 2 November 1953.

Citation: Sgt. Bleak, a member of the medical company, distinguished himself by conspicuous gallantry and indomitable courage above and beyond the call of duty in action against the enemy. As a medical aidman, he volunteered to accompany a reconnaissance patrol committed to engage the enemy and capture a prisoner for interrogation. Forging up the rugged slope of the key terrain, the group was subjected to intense automatic weapons and small arms fire, and suffered several casualties. After administering to the wounded, he continued to advance with the patrol. Nearing the military crest of the hill, while attempting to cross the fire-swept area to attend the wounded, he came under hostile fire from a small group of the enemy concealed in a trench. Entering the trench, he closed with the enemy, killed two with bare hands and a third with his trench knife. Moving from the emplacement, he saw a concussion grenade fall in front of a companion and, quickly shifting his position, shielded the man from the impact of the blast. Later, while ministering to the wounded, he was struck by a hostile bullet, but, despite the wound, he undertook to evacuate a wounded comrade. As he moved down the hill with his heavy burden, he was attacked by two enemy soldiers with fixed bayonets. Closing with the aggressors, he grabbed them and smacked their heads together, then carried his helpless comrade down the hill to safety. Sgt. Bleak's dauntless courage and intrepid actions reflect utmost credit upon himself and are in keeping with the honored traditions of the military service.

### Collier, Gilbert G.

Rank and organization: Sergeant (then Cpl.), U.S. Army, Company F, 223d Infantry Regiment, 40th Infantry Division. Place and date: Near Tutayon, Korea, 19–20 July 1953. Entered service at: Tichnor AK. Born: 30 December 1930, Hunter, AK. G.O. No.: 3, 12 January 1955.

Citation: Sgt. Collier, a member of Company F, distinguished himself by conspicuous gallantry and indomitable courage above and beyond the call of duty in action against the enemy. Sgt. Collier was pointman and assistant leader of a combat patrol committed to make contact with the enemy. As the patrol moved forward through the darkness, he and his commanding officer slipped and fell from a steep, 60-foot cliff and were injured. Incapacitated by a badly sprained ankle which prevented immediate movement, the officer ordered the patrol to return to the safety of friendly lines. Although suffering from a painful back injury, Sgt. Collier elected to remain with his leader, and before daylight they managed to crawl back up and over the mountainous terrain to the opposite valley where they concealed themselves in the brush until night-fall, then edged toward their company positions. Shortly after leaving the daylight retreat they were ambushed and, in the ensuing fire fight, Sgt. Collier killed two hostile soldiers, received painful wounds, and was separated from his companion. Then, ammunition expended, he closed in hand-to-hand combat with four attacking hostile infantrymen, killing, wounding, and routing the foe with his bayonet. He was mortally wounded during this action but made a valiant attempt to reach and assist his leader in a desperate effort to save his comrade's life without regard for his own personal safety. Sgt. Collier's unflinching courage, consummate devotion to duty, and gallant self-sacrifice reflect lasting glory upon himself and uphold the noble traditions of the military service.

## Hernandez, Rodolfo P.

Rank and organization: Corporal, U.S. Army, Company G, 187th Airborne Regimental Combat Team. Place and date: Near Wontong-ni, Korea, 31 May 1951. Entered service at: Fowler, CA. Born: 14 April 1931, Colton, CA. G.O. No.: 40, 21 April 1952.

**Citation**: Cpl. Hernandez, a member of Company G, distinguished himself by conspicuous gallantry and intrepidity above and beyond the call of duty in action against the enemy. His platoon, in defensive positions on Hill 420, came under ruthless attack by a numerically superior and fanatical hostile force, accompanied by heavy artillery, mortar, and machine gun fire which inflicted numerous casualties on the platoon. His comrades were forced to withdraw due to lack of ammunition, but Cpl. Hernandez, although wounded in an exchange of grenades, continued to deliver deadly fire into the ranks of the onrushing assailants until a ruptured cartridge rendered his rifle inoperative. Immediately leaving his position, Cpl. Hernandez rushed the enemy armed only with rifle and bayonet. Fearlessly engaging the foe, he killed six of the enemy before falling unconscious from grenade, bayonet, and bullet wounds, but his heroic action momentarily halted the enemy advance and enabled his unit to counterattack and retake the lost ground. The indomitable fighting spirit, outstanding courage, and tenacious devotion to duty clearly demonstrated by Cpl. Hernandez reflect the highest credit upon himself, the infantry, and the U.S. Army.

## Speicher, Clifton T.

Rank and organization: Corporal, U.S. Army, Company F, 223d Infantry Regiment, 40th Infantry Division. Place and date: Near Minarigol, Korea, 14 June 1952. Entered service at: Gray, PA. Born: 25 March 1931, Gray, PA. G.O. No.: 65, 19 August 1953.

**Citation**: Cpl. Speicher distinguished himself by conspicuous gallantry and indomitable courage above and beyond the call of duty in action against the enemy. While participating in an assault to secure a key terrain feature, Cpl. Speicher's squad was pinned down by withering small-arms, mortar, and machine gun fire. Although already wounded, he left the comparative safety of his position and made a daring charge against the machine gun emplacement. Within 10 yards of the goal he was again wounded by small-arms fire but continued on, entered the bunker, killed two hostile soldiers with his rifle, a third with his bayonet, and silenced the machine gun. Inspired by this incredible display of valor, the men quickly moved up and completed the mission. Dazed and shaken, he walked to the foot of the hill where he collapsed and died. Cpl. Speicher's consummate sacrifice and unflinching devotion to duty reflect lasting glory upon himself and uphold the noble traditions of the military service.

# APPENDIX F:
# HOYT S. VANDENBERG,
# 1899–1954

Vandenberg Air Force Base is named in honor of the late General Hoyt Sanford Vandenberg, second Air Force Chief of Staff of the United States Air Force, and chief architect of today's modern Air Force.

Hoyt Vandenberg was born in Milwaukee, Wisconsin, on January 24, 1899. In 1923 he graduated from West Point Academy, ranking 240 in a class of 261. Vandenberg excelled in pilot training at both Brooks and Kelly Field in Texas. He flew attack and fighter aircraft, and served two tours as an instructor pilot. His reputation as an outstanding pilot earned him a series of education assignments at the Air Corps Tactical School, Maxwell AFB, Alabama; the Command and General Staff College, Fort Leavenworth, Kansas; and the Army War College, Washington, D.C.

In June 1939 he was assigned to the plans division of the chief of the Air Corps. After the United States had entered World War II he was appointed operations and training officer of the Air Staff under Gen. Henry H. (Hap) Arnold. During the early stages of the war, Vandenberg (then a colonel) was transferred to England and assisted in planning air operations for the invasion of North Africa. He received his first star in December 1942 and became chief of staff of the Twelfth Air Force in North Africa under Gen. James H. Doolittle. During this campaign he flew more than two dozen combat missions over Tunisia, Italy, Sardinia, Sicily, and Panteileria to obtain firsthand information.

Returning to the United States in August 1943, Gen. Vandenberg was assigned to Army Air Force Headquarters as deputy chief of the Air Staff. A month later he became head of an Air Mission to Russia under Ambassador W. Averell Harriman, and returned to the United States in January 1944. In

March he was promoted to major general and returned to Europe as deputy air commander-in-chief of the Allied Expeditionary Forces and commanding general of its American air component. He helped plan the Normandy invasion, and in August 1944 took over command of the Ninth Air Force in the European Theater.

In March 1945 he was promoted to the rank of lieutenant general, and full general in 1947. Meanwhile, in January 1946, Gen. Vandenberg was appointed chief of the intelligence division of the General Staff. In June he was named director of the Central Intelligence Group, predecessor to the Central Intelligence Agency, formed in 1947.

**Gen. Hoyt S. Vandenberg. U.S. Air Force.**

He returned to duty with the Air Force in May 1947 and became deputy commander and chief of staff of the Army Air Force. With the establishment of a separate Air Force in September 1947, Vandenberg became its first vice chief of staff under Gen. Carl Spaatz, and succeeded him on April 30, 1948. He held that post through the critical periods of the Berlin airlift (1948–49) and the Korean War (1950–53).

Weak, exhausted, and in constant pain from cancer, Gen. Vandenberg retired from the Air Force in June 1953. He died in Washington, D.C. on April 2, 1954. In honor of his service to the nation, the aerospace base near Lompoc, California, formerly Cooke Air Force Base, was renamed Vandenberg Air Force Base on October 4, 1958.

Source: Vandenberg Air Force Base Internet page. www.Vandenberg.af.mil/library/factsheet_print.asp?fsID=4606 (accessed May 3, 2013).

# APPENDIX G:
# VANDENBERG AFB LAUNCH
# FACILITY STATUS AND HISTORY

| Launch Facility & Bldg. Number | Acceptance Date by Air Force* | Date of First Launch | Present Status | Original Purpose | Current or (Last) Use | Former Designation | Remarks |
|---|---|---|---|---|---|---|---|
| SLC-2E 1620 (soft) | Jul 1958 | 16 Dec 1958 | Decommissioned & stripped | Thor IRBM | (Thor-Delta) (space) | 75-1-1 | *First Vandenberg missile launch.*† First space launch, 16 Jun 1961. Last launch, 11 Mar 1972. Total launches: 51. Transferred to NASA with SLC-2W. |
| SLC-2W 1623 (soft) | Jul 1958 | 17 Sep 1959 | Active (NASA operated) | Thor IRBM | (Thor-Delta) (space) | 75-1-2 | First space launch, 28 Aug 1962. Facility modified (1992–1993) for Delta II space boosters. |
| SLC-1E 1642 (soft) | Oct 1958 | 25 Jun 1959 | Decommissioned & stripped | Thor-Agena (space) | Thorad-Agena (space) | 75-3-5 | Last launch, 18 Sep 1968. Total launches: 45. |
| SLC-1W 1635 (soft) | Oct 1958 | *28 Feb 1959* | Decommissioned & stripped | Thor-Agena (space) | Thorad-Agena (space) | 75-3-4 | *First Vandenberg space launch* and the world's first polar orbiting satellite. Last launch, 14 Dec 1971. Total launches: 55. |
| SLC-10E 1651 (soft) | Oct 1958 | 16 Jun 1959 | Decommissioned & stripped | Thor IRBM | (Thor) (training) | 75-2-7 LE-7 | (RAF CTL program) Last launch, 19 Mar 1962. Total launches: 6. |

*Acceptance of "brick and mortar" prior to installation and checkout of aerospace equipment. †Italicized launch dates correspond with italicized remarks.

| Launch Facility & Bldg. Number | Acceptance Date by Air Force* | Date of First Launch | Present Status | Original Purpose | Current or (Last) Use | Former Designation | Remarks |
|---|---|---|---|---|---|---|---|
| SLC-10W 1658 (soft) | Oct 1958 | 14 Aug 1959 | National Historic Landmark | Thor IRBM | (Thor-Block 5D) (space) | 75-2-6 LE-6 4300 B-6 | Space launch facility for DMSP. Last launch, 14 Jul 1980. Total launches: 20. Designated National Historic Landmark by U.S. Department of Interior, 18 Jul 1986. |
| LE-8 1661 (soft) | Oct 1958 | 16 Apr 1959 | Decommissioned & stripped | Thor IRBM | (Thor IRBM) (RAF CTL) | 75-2-8 | First weapon system training launch RAF-Thor IWST program. Final RAF and Thor IRBM launch, 18 Jun 1962. Total launches: 7. |
| BMRS A-1 1797 (soft) | May 1958 | 26 Oct 1962 | Removed | Atlas D ICBM | (Atlas F) (space) | 65-1-1 576 A-1 4300 A-1 ABRES A-1 | First U.S. nuclear ICBM on alert, 31 Oct 1959. Site of first ABRES launch, 4 Nov 1963. Last launch, 8 Sep 1974. Total launches: 36. MST and gantry removed in 1984. |
| BMRS A-2 1790 (soft) | Nov 1958 | 9 Sep 1959 | Removed | Atlas D ICBM | (Atlas F) (space) | 65-1-2 576 A-2 4300 A-2 ABRES A-2 | First Vandenberg ICBM launch. First ABRES from this site, 5 Aug 1965. First space launch, 6 Apr 1968. Last launch, 6 Aug 1971. Total launches: 14. Gantry and MST removed, 20 Apr 1984. |

| Launch Facility & Bldg. Number | Acceptance Date by Air Force* | Date of First Launch | Present Status | Original Purpose | Current or (Last) Use | Former Designation | Remarks |
|---|---|---|---|---|---|---|---|
| BMRS A-3 (soft) 1788 | Apr 1959 | 26 Jan 1960 | Decommissioned & stripped | Atlas D ICBM | (Commercial Space) | 65-1-3 576 A-3 4300 A-3 ABRES A-3 | Site modified in 1989 to launch first commercial space booster from Vandenberg by the American Rocket Company. First launch attempt, 5 Oct 1989, a failure. No further attempts. Total launches: 34. |
| ABRES B-1 1835 Coffin-type (semi-hard) | Jun 1959 | 22 Jul 1960 | Decommissioned & stripped | Atlas D ICBM | (Atlas D) (ABRES program) | 65-2-1 576B-1 | First Atlas D off alert, 1 May 1964. First ABRES launch from this site, 6 Apr 1965. Last launch, 10 Jun 1966. Total launches: 13. |
| ABRES B-2 1825 Coffin-type (semi-hard) | Jun 1959 | 22 Apr 1960 | Decommissioned & stripped | Atlas D ICBM | (Atlas D) (ABRES program) | 65-2-2 576 B-2 | First ABRES launch from this site, 3 Jun 1965. Last launch, 7 Nov 1967. Total launches: 27. |
| ABRES B-3 1820 Coffin-type (semi-hard) | Aug 1959 | 12 Sep 1960 | Decommissioned & stripped | Atlas D ICBM | (Atlas D) (ABRES program) | 65-2-3 576 B-3 | First ABRES launch, 21 Jan 1965. First space launch, 27 May 1965. Last launch, 11 Oct 1967. Total launches: 21. |

| Launch Facility & Bldg. Number | Acceptance Date by Air Force* | Date of First Launch | Present Status | Original Purpose | Current or (Last) Use | Former Designation | Remarks |
|---|---|---|---|---|---|---|---|
| SLC-3W 770 (soft) | Sep 1959 | 11 Oct 1960 | Removed | Atlas-Agena Space | (Atlas E) (Space) | PALC 1-1 | Converted from Atlas to Thor (space) in 1963. Back to Atlas space booster in 1973. Last launch, 24 Mar 1995. Total launches: 81. Leased to Lockheed Martin for Atlas V, 1 Apr 1999. Removed MST, 22 Jan 2000. Lease terminated 1 Aug 2003. No launches. Leased to SpaceX, 4 Feb 2004. Suspended operations late 2005, terminated lease 31 Dec 2011. No launches. |
| SLC-3E 751 (soft) | Sep 1959 | 12 Jul 1961 | Replaced | Atlas Space | (Atlas AS) Atlas V Space | PALC 1-2 | Site modified in 1983 for Atlas H. Last launch from original facility, 15 May 1987. Total launches: 18. MST replaced for Atlas IIAS. Three launches, last on 2 Dec 2003. Leased to Lockheed Martin for Atlas V, 1 Jul 2004. Facility modified. First launch, 13 Mar 2008. |

| Launch Facility & Bldg. Number | Acceptance Date by Air Force* | Date of First Launch | Present Status | Original Purpose | Current or (Last) Use | Former Designation | Remarks |
|---|---|---|---|---|---|---|---|
| OSTF 1861 silo-lift (hard) | Dec 1959 | None | A scarred hole in the ground | Titan I Test Facility | (Weapon System Demonstration) | None | Destroyed 3 Dec 1960 by in-silo deflagulation during a simul. launch. No launches. |
| PLC-A 939 (soft) | Dec 1959 | 14 Jul 1959 | Decommissioned & stripped | Probe vehicles (space) | (Nike-Javelin) | NERV-Sunflare Pad A PALC-A | Previous Scout Jr. launcher. Constructed for AEC. Last launch, 25 Mar 1966. Total launches: 39. |
| PALC-B none (soft) | Dec 1959 | 4 Feb 1960 | Decommissioned & stripped | Probe vehicles (space) | (Kiva-Hopi) (space probe) | Tumbleweed Pad B | Constructed for AEC. Last launch, 11 May 1963. Total launches: 23. |
| 576-G 1935 Silo-lift (hard) | Mar 1960 | 10 Aug 1962 | Decommissioned & stripped | Atlas F ICBM | (Atlas F) (special test) | OSTF-2 | Last launch, 8 Jan 1965. Total launches: 7. |
| 576-F 1836 Coffin-type (semi-hard) | May 1960 | 7 Jun 1961 | Decommissioned & stripped | Atlas E ICBM | (Atlas E) (Nike X) | OSTF-1 | Last launch, 27 Aug 1964. Total launches: 10. |
| 576-C 1895 Coffin-type (semi-hard) | Jul 1960 | 3 Jul 1963 | Decommissioned & stripped | Atlas E ICBM | (Atlas E) (DASO) | None | Last launch, 25 Sep 1963. Total launches: 3. |
| SLTF 1885 Silo (hard) | Aug 1960 | 3 May 1961 | Decommissioned & stripped | Titan I/II Silo test | Titan I/II Silo test | None | *First Vandenberg in-silo launch.* A Titan I launched from a prototype Titan II silo to determine the environmental effect on the emplaced missile and facility during an in-silo liftoff. Total launches: 1. |

| Launch Facility & Bldg. Number | Acceptance Date by Air Force* | Date of First Launch | Present Status | Original Purpose | Current or (Last) Use | Former Designation | Remarks |
|---|---|---|---|---|---|---|---|
| 395-A (LE-1) 1875 Silo-lift (hard) | Aug 1960 | 23 Sep 1961 | Decommissioned & stripped | Titan I ICBM | (Titan I) (special test) | None | Last launch, 8 Dec 1964. Total launches: 11. |
| 395-A (LE-2) 1877 Silo-lift (hard) | Aug 1960 | 30 Mar 1963 | Decommissioned & stripped | Titan I ICBM | Titan I (special test) | None | Last launch 5 Mar 1965. Total launches: 4. |
| 395-A (LE-3) 1879 Silo-lift (hard) | Sep 1960 | 20 Jan 1962 | Decommissioned & stripped | Titan I ICBM | Titan I (special test) | None | Last launch, 14 Jan 1965. Total launches: 4. |
| 395-B 1799 Silo (hard) | Jun 1961 | 17 Feb 1964 | Decommissioned & stripped | Titan II ICBM | Titan II (OT) | None | Last launch on 20 May 1969 involved an ICBM that had been on strategic alert in this silo for a year. Total launches: 16. |
| 395-C 1050 Silo (hard) | May 1961 | *16 Feb 1963* | Decommissioned & partially stripped | Titan II ICBM | Titan II (special test) | None | *First Vandenberg Titan II launch.* Last launch, 27 Jun 1976. Total launches: 31. |
| 395-D 1520 Silo (hard) | May 1961 | 13 May 1963 | Decommissioned & stripped | Titan II ICBM | Titan II (OT) | None | Last launch, 5 Apr 1966. Total launches: 11. |
| 576-D 1920 Silo-lift (hard) | Jun 1961 | 15 May 1963 | Decommissioned & stripped | Atlas F ICBM | Atlas F (DASO) | None | Last launch, 31 Aug 1964. Total launches: 2. |
| 576-E 1611 Silo-lift (hard) | Jul 1961 | 1 Aug 1962 | Silo: Decommissioned & stripped | Atlas F ICBM | (Atlas F) (Special test) | None | Last Launch, 22 Dec 1964. Total launches: 4. |

| Launch Facility & Bldg. Number | Acceptance Date by Air Force* | Date of First Launch | Present Status | Original Purpose | Current or (Last) Use | Former Designation | Remarks |
|---|---|---|---|---|---|---|---|
| Concrete launch pad adjacent to 576-E silo (No facility number) | May 1993 | 13 May 1994 | Launch pad: Active | Taurus Space | Taurus Space | None | Adjacent to the sealed silo, a concrete apron and launch stand were built for the Taurus space booster. First launch, 13 Mar 1994. |
| SLC-5 580 (soft) | Nov 1961 | 26 Apr 1962 | Removed | Scout Space | (Scout Space) | Pad D PALC-D | NASA/European space satellites launched aboard American boosters. Last launch, 8 May 1994. Total launches: 69. All facilities removed, 11 Apr 2011. |
| LF-02 1971 Silo (hard) | Nov 1961 | 12 Apr 1963 | Active | Minute-man I ICBM | (Peacekeeper ICBM) Current use: Boost Vehicle | 394 A-1 | Converted to Peacekeeper, Sep 1985. First Peacekeeper launch, 23 Aug 1986. Last launch, 12 Mar 2003. Transferred to Missile Defense Agency, Mar 2003 for Ground-based Midcourse Defense (GMD) program. Silo converted for Boost Vehicle (interceptor) missiles, Aug 2004. |

| Launch Facility & Bldg. Number | Acceptance Date by Air Force* | Date of First Launch | Present Status | Original Purpose | Current or (Last) Use | Former Designation | Remarks |
|---|---|---|---|---|---|---|---|
| LF-03　1972 Silo (hard) | Nov 1961 | 30 Apr 1963 | Active | Minuteman I ICBM | (Minuteman II) (GMD MSLS)<br><br>Current use: Boost Vehicle | 394 A-2 | Last launch, 14 Jul 2001. Transferred to Missile Defense Agency, Mar 2003 for Ground-based Midcourse Defense (GMD) program. Silo converted for Boost Vehicle (interceptor) missiles, Sep 2004. |
| LF-04　1976 Silo (hard) | Nov 1961 | 28 Sep 1962 | Active | Minuteman I ICBM | Minuteman III ICBM | 394 A-3 | *First Vandenberg Minuteman* launch. Site operated by 576th Flight Test Squadron (AFGSC). |
| LF-05　1977 Silo (hard) | Nov 1961 | 10 Dec 1962 | Caretaker status | Minuteman I ICBM | (Peacekeeper ICBM) | 394 A-4 | Converted to Peacekeeper, Feb 1985. Last launch, 21 Jul 2004. Total launches: 60. Turned over to 30th Civil Engineer Squadron (AFSPC), facility badly deteriorated. Scheduled to be filled in. |
| LF-06　1980 Silo (hard) | Dec 1961 | 11 Apr 1963 | Caretaker status | Minuteman I ICBM | (Minuteman II ICBM) | 394 A-5 | Modified to Minuteman III, 1987; modified in 2004 for Minuteman II TLV program. Last launch, 23 Sep 2008. Total launches: 85. Turned over to 30th Civil Engineer Squadron (AFSPC), 27 Feb 2012. Scheduled to be filled in. |

| Launch Facility & Bldg. Number | Acceptance Date by Air Force* | Date of First Launch | Present Status | Original Purpose | Current or (Last) Use | Former Designation | Remarks |
|---|---|---|---|---|---|---|---|
| PLC-C 589 (soft) | 30 Mar 1962 | 29 Jun 1971 | Decommissioned & stripped | Probe vehicles space | (UTE Tomahawk) (space probe) | Pad C PALC-C | *Placed on the roof of the Scout blockhouse.* Not used at first because of interference with nearby Pad D (SLC-5). Last launch, 10 Dec 1975. Total launches: 6. |
| LF-07 1981 Silo (hard) | Apr 1962 | 24 May 1963 | Caretaker status | Minuteman I ICBM | (Minuteman II ICBM) (OT) | 394 A-6 | Last launch, 9 Nov 1987. Total launches: 64. Facility badly deteriorated. |
| SLC-4E 715 (soft) | Nov 1962 | 14 Aug 1964 | Removed | Atlas-Agena Space | (Titan IV) (Space) | PALC-2-4 | Converted to Titan IIID facility in 1971. Modified in 1983 to also accommodate Titan 34D. Converted to Titan IV facility in 1989. First Titan IV launch, 8 Mar 1991; last launch, 19 Oct 2005. Total Titan launches: 41. SLC-4E licensed to SpaceX on 8 Jun 2012. Site demolition began in May 2011 for new launch facility to support Falcon 9 rocket. First launch, 29 Sep 2013. |

| Launch Facility & Bldg. Number | Acceptance Date by Air Force* | Date of First Launch | Present Status | Original Purpose | Current or (Last) Use | Former Designation | Remarks |
|---|---|---|---|---|---|---|---|
| SLC-4W 738 (soft) | Nov 1962 | 12 Jul 1963 | Decommissioned & stripped | Atlas-Agena Space | (Titan II SLV) (Space) | PALC-2-3 | Converted to Titan IIIB facility in 1966. Modified to Titan II SLV facility. First Titan II SLV launch, 5 Sep 1988. Last launch, 18 Oct 2003. Total Titan launches: 81. |
| 4300-C 1681 (soft) | Apr 1963 | 17 Dec 1963 | Decommissioned & stripped | Scout Junior Space | (Scramjet) (space) | 279L | First SAC space launch, a Scout Jr. by the 4300th Support Squadron (4300th Support Group). Last launch, 11 Jan 1967. Total launches: 2. |
| LF-08 1986 Silo (hard) | Mar 1963 | 26 Sep 1963 | Trainer facility | Minute-man I ICBM | (Peacekeeper) ICBM | 394 A-7 | Converted to Peacekeeper, Mar 1985. First Peacekeeper in-silo launch, 23 Aug 1985. Last launch, 11 Jun 1991. Total launches: 66. Site maintained by 532nd Training Squadron (AETC). |
| LF-09 1993 Silo (hard) | Jan 1964 | 29 Jun 1964 | Active | Minute-man I ICBM | Minuteman III ICBM (FDE) | 394 A-8 | Site operated by 576th Flight Test Squadron (AFGSC). |

| Launch Facility & Bldg. Number | Acceptance Date by Air Force* | Date of First Launch | Present Status | Original Purpose | Current or (Last) Use | Former Designation | Remarks |
|---|---|---|---|---|---|---|---|
| LF-21   1962 Silo (hard) | Aug 1964 | 18 Aug 1965 | Active | Minuteman II ICBM | (Minuteman II/III ICBM) Current use: Boost Vehicle | None | Transferred to Missile Defense Agency in Mar 2003 for GMD program. Silo converted for Boost Vehicle (interceptor) missiles, Oct 2004. |
| LF-10   1963 Silo (hard) | Nov 1964 | 6 Oct 1965 | Active | Minuteman II ICBM | Minuteman III ICBM (FDE) | LF-22 | Site operated by 576th Flight Test Squadron (AFGSC). First launch as LF-10, 12 Jul 1987. |
| LF-23   1964 Silo (hard) | Oct 1964 | 26 Aug 1966 | Active | Minuteman II ICBM | Boost Vehicle | None | Only one Minuteman launch from this facility. Facility transferred to Missile Defense Agency and modified for Boost Vehicle (interceptor) missiles in the Ground-based Midcourse Defense (GMD) program. First operational launch of BV from Vandenberg, 1 Sep 2006. |
| LF-24   1965 Silo (hard) | Jan 1965 | 15 Dec 1965 | Active | Minuteman II ICBM | (Minuteman III ICBM) Current use: Boost Vehicle | None | Transferred to Missile Defense Agency for use in GMD program. Silo converted for Boost Vehicle (interceptor) missiles. |

| Launch Facility & Bldg. Number | Acceptance Date by Air Force* | Date of First Launch | Present Status | Original Purpose | Current or (Last) Use | Former Designation | Remarks |
|---|---|---|---|---|---|---|---|
| LF-25 1966 Silo (hard) | Jan 1965 | 16 Feb 1966 | Decommissioned & stripped | Minuteman II ICBM | (Minuteman III ICBM) | None | Total launches: 22. Last launch, 4 Mar 1976. Facility badly deteriorated. Scheduled to be filled in. |
| LF-26 1967 Silo (hard) | Feb 1965 | 18 Jan 1966 | Caretaker status | Minuteman II ICBM | (Minuteman III ICBM) | None | Last launch, 7 Apr 2006. Total launches: 66. Site maintained by 576 Flight Test Squadron (AFGSC) |
| Bomarc-1 1815 (soft) | Feb 1966 | 14 Oct 1966 | Decommissioned | Bomarc A Target | (Bomarc B) (U.S. Navy Target) | None | For fleet SAM/AAM target practice. Last launch, 14 Jul 1982. Total launches: 49. |
| Bomarc-2 1817 (soft) | Feb 1966 | 25 Aug 1966 | Decommissioned | Bomarc A Target | (Bomarc B) (U.S. Navy Target) | None | For fleet SAM/AAM target practice. Last launch, 14 Jul 1982. Total launches: 38. |
| SLC-6 390 (soft) | Feb 1970 (MOL) May 1984 (STS) | None | Active. Leased site | Titan IIIM (Space MOL) | (Space Shuttle) (No launches) (LLV) (Athena) Delta IV | None | Converted for Space Shuttle use (1979–1986). Assigned to mothball status, May 1988. Planned modification for Titan IV-Centaur (1987–1991). Non-exclusive 5-year lease to Lockheed (30 Dec 1994). First rocket launch, LLV-1, 15 Aug 1995. Leased to Boeing, 1 Sep 1999. Major modifications for Delta IV. First launch, 27 Jun 2006. |

| Launch Facility & Bldg. Number | Acceptance Date by Air Force* | Date of First Launch | Present Status | Original Purpose | Current or (Last) Use | Former Designation | Remarks |
|---|---|---|---|---|---|---|---|
| TP-01 (soft) 1840 | Mar 1982 | 17 Jun 1983 | Decommissioned & stripped | Peacekeeper ICBM (Canister) | (Small ICBM) (Canister) | None | First Vandenberg "cold launch" of Peacekeeper, 17 Jun 1983. First Vandenberg "cold launch" of Small ICBM, 11 May 1989. Latter program cancelled, Jan 1992. Modified for Conventional Strike Missile, 2013. |
| RGLS (soft) 1862 | Late 1990 | None | Decommissioned & stripped Garrison | Peacekeeper Rail | Peacekeeper (Peacekeeper Rail Garrison) —no launches | None | Consisted of multiple facilities and 6.5 miles of railroad track. (1988–1990). Pres. George Bush cancelled RGLS, 27 Sep 1991. No testing or launches from the site. |
| SLC-8 (Soft) 240 | Privately built and operated launch facility. | 11 Apr 2005 | Active | Minotaur I Space | Minotaur I and IV | SLF | WCSC signed 25-year lease with Air Force for approx. 104.9 acres of land south of SLC-6 to construct a commercial spaceport on the prop. to launch government and commercial vehicles. Agreement signed 16 Mar 1995, eff. retroactively to 6 Mar. Company reorganized into SSI. With Air Force approval, renamed launch facility SLC-8 on 1 Dec 2003. |

## *Terms*

| | |
|---|---|
| *Soft* | Missile emplaced and launched from exposed launcher above ground (gantry-type or test pad). |
| *Semi-hard* | Missile emplaced in protective ground-level concrete housing prior to launch (see coffin-type, below). |
| *Hard* | Missile emplaced in protective ground-level concrete housing prior to launch (see silo and silo-lift, below). |
| *Silo* | Protective vertical tube-type housing from which a missile is launched from in emplaced position (Minuteman). |
| *Silo-lift* | Missile is stored in silo, but is elevated clear of the tube for above-ground launching (Titan I). |
| *Coffin-Type* | Enclosed horizontal storage from which the missile is raised to exposed vertical launch position (Thor, Atlas D). |
| *Cold Launch* | Missile is ejected from a tube by gas pressure, and ignites Stage 1 propellant after it is airborne (Peacekeeper, Small ICBM). |
| *Stripped* | All recoverable material sold and removed under civilian contract, or retained by the Air Force, but transferred to other organizations. |

## *Glossary*

| | |
|---|---|
| **AAM** | Air-to-Air Missile |
| **ABRES** | Advanced Ballistic Reentry System |
| **AEC** | Atomic Energy Commission |
| **AETC** | Air Education and Training Command |
| **AFGSC** | Air Force Global Strike Command |
| **AFSPC** | Air Force Space Command |
| **BMRS** | Ballistic Missile Reentry System |
| **BV** | Boost Vehicle |
| **CTL** | Combat Training Launch |
| **DASO** | Demonstration and Shakedown Operations |
| **FDE** | Force Development Evaluation |
| **GMD** | Ground-based Midcourse Defense |
| **ICBM** | Intercontinental Ballistic Missile |
| **IRBM** | Intermediate Range Ballistic Missile |
| **IWST** | Integrated Weapon System Training |
| **LE** | Launch Emplacement |
| **LF** | Launch Facility |
| **MOL** | Manned Orbiting Laboratory |
| **MSLS** | Multi-Service Launch System |
| **MST** | Mobile Service Tower |
| **OSTF** | Operational System Test Facility |
| **OT** | Operational Test |
| **PALC** | Point Arguello Launch Complex |
| **PLC** | Probe Launch Complex |
| **RAF** | Royal Air Force (United Kingdom) |
| **RGLS** | Rail Garrison Launch Site (Peacekeeper) |
| **SAC** | Strategic Air Command |
| **SAM** | Surface-to-Air Missile |

| | |
|---|---|
| **SLC** | Space Launch Complex |
| **SLTF** | Silo Launch Test Facility |
| **SLV** | Space Launch Vehicle |
| **SSI** | Spaceport Systems International |
| **STS** | Space Transportation System (Space Shuttle) |
| **WCSC** | Western Commercial Space Center |

Data compiled by the author from multiple documents at the 30th Space Wing History Office, Vandenberg AFB, CA.

# CHAPTER NOTES

## Introduction

1. Margaret E. Wagner, Linda B. Osborn, Susan Reyburn, and the Staff of the Library of Congress, eds., *World War II Companion* (New York: Simon & Schuster, 2007), 88.

2. "Original Jesus Maria Rancho Is Offered to Military Unit," *Lompoc Record*, March 7, 1941, 1; "Camp Survey Job Here Awarded," *Santa Maria Daily Times*, May 22, 1941, 1, 4; Averam B. Bender, "From Tanks to Missiles: Camp Cooke/Cooke Air Force Base (California), 1941–1958," *Arizona and the West*, Autumn 1967, 221.

3. "Camp Beginning Hailed by Talks as Dirt Thrown," *Santa Maria Daily Times*, September 15, 1941, 1–2; Bender, 220; Sgt. Wesley W. Purkiss, ed., "A History of Camp Cooke, 1941 to 1946," April 1946, 6, 22–23.

4. First Strategic Aerospace Division History Office, "The Story of Marshallia Ranch: A History Dating Back to the Olivera Adobe Built in 1837," historical vignette, Vandenberg AFB, Ca., n.d., 1–2, 4–5.

5. Shelby L. Stanton, *Order of Battle U.S. Army, World War II* (Novato, California: Presidio Press, 1984), 54, 55, 63, 66, 69, 158, 173, and 268.

6. "'This Is the Army' Movie Scenes Filmed at Camp; Joe Louis Here," *Camp Cooke Clarion*, March 5, 1943, 5; "Joe Louis Exhibition Only Fair' 'Sugar' Robinson Cops Spotlight," *Cooke Clarion*, November 19, 1943, 4–5.

7. "Entertainment: USO Camp Shows," http://www.uso.org/whatwedo/entertainment/historicaluso campshows (accessed June 8, 2010).

8. Ibid.

9. Ibid.; booklet, Hollywood Victory Committee, *HVC*, circa 1945, n.p., "United Service Organizations," Margaret Herrick Library, Academy of Motion Picture Arts and Sciences, Beverly Hills, Ca.

10. "Big Camel Stage Show, Arena Next Tuesday," *Camp Cooke Clarion*, October 23, 1942, 1, 3; "'Shell Show,' Variety Hit, to Play Here Two

Nights," *Cooke Clarion*, January 21, 1944, 1, 3; "'Clambake Follies' Plays to Packed Arena," *Cooke Clarion*, June 23, 1944, 1, 3.

11. "'This Is the Army' Movie Scenes Filmed at Camp; Joe Louis Here," *Camp Cooke Clarion*, March 5, 1943, 5; "20th Century–Fox Using Division in Filming 'I Married a Soldier,'" *Camp Cooke Clarion*, December 10, 1943, 1, 6; "Tankers, Depotmen in Film of Stalingrad Fight," *Camp Cooke Clarion*, December 1, 1944, 1, 3.

12. T/5 Holzer, "212th Throws 'Camp House Party,'" *Cooke Clarion*, August 13, 1943, 1, 6; Pvt. Paul Givin, "156th Engineers Say Two-Toned Whistle Best Describes Weekend," *Cooke Clarion*, March 17, 1944, 4.

13. "Marjorie Hall Bids Adieu to Friends in Camp," *Cooke Clarion*, July 7, 1944, 6.

14. "Mrs. Stockton Takes Over Red Cross Position," *Cooke Clarion*, January 5, 1945, 4; "Six-Act Variety Revue Presented for Hospital GIs," *Cooke Clarion*, October 20, 1944, 3.

15. "Fun Programs of Mrs. Faye Porter May Be Ended by War Rationing," *Cooke Clarion*, November 27, 1942, 2; "AWVS Plans Christmas Entertainment for Armed Forces at Center and Camp," *Lompoc Record*, December 20, 1951, 1.

16. "Obituary: Faye Porter," *Lompoc Record*, April 5, 1974, 2.

17. "Work of Italian Service Units Here Described," *Cooke Clarion*, April 6, 1945, 1, 3.

18. Purkiss, 82.

19. "USDB Activation to Be Immediate," *Lompoc Record*, December 12, 1946, 1.

20. "Call 40th Into Federal Service," *Santa Maria Times*, August 1, 1950, 1, 3; "First 44th Unit Bound for Lewis in Motor Convoy," *Santa Maria Times*, November 26, 1952, 1.

21. "466th Battalion Makes Self at Home at Camp Cooke; Completing Training," *Lompoc Record*, January 24, 1952, II-1.

22. "Annual Report of Army Medical Services Activities 1951: United States Army Hospital Specialized Medical Treatment Center, Camp Cooke,

California," January 15, 1952, 19, AGF and AFF Installations, Camp Cooke, Records of Headquarters Army Ground Forces, RG 337, Box 16, National Archives and Records Administration, College Park, MD.

23. "Headquarters Camp Cooke, California: Official Diary," n.d., 137, AGF and AFF Installations, Camp Cooke, Records of Headquarters Army Ground Forces, RG 337, Box 17, National Archives and Records Administration, College Park, MD.

24. Jeffrey Geiger, "Vandenberg Launch Summary," n.d., History Office, 30th Space Wing, Vandenberg AFB, CA, hereafter cited as SWHO.

## Chapter 1

1. "Original Jesus Maria Rancho Is Offered to Military Unit," *Lompoc Record*, March 7, 1941, 1; "Option Filing Indicates Interest in Jesus Maria for Huge Military Site," *Lompoc Record*, March 21, 1941, 1; "Rumors Persist That U.S. Army Plans Mesa Camp," *Lompoc Record*, March 28, 1941,1; "Army Camp Plans Call for Military City to Spring Up in Five Months," *Santa Barbara News-Press*, August 3, 1941, 15, 20; "Huge Crew to Start Giant Army Camp Job at Once," *Santa Barbara News-Press*, September 7, 1941, 13, 18.

2. Sgt. Wesley W. Purkiss, ed., "A History of Camp Cooke, 1941 to 1946," April 1946, 2; Averam B. Bender, "From Tanks to Missiles: Camp Cooke/Cooke Air Force Base (California), 1941–1958," *Arizona and the West*, Autumn 1967, 221; "U.S. Sues for Camp Cooke Lands," *Santa Maria Daily Times*, April 18, 1842, 1.

3. "Camp Beginning Hailed by Talks as Dirt Thrown," *Santa Maria Daily Times*, September 15, 1941, 1–2; "Work on Huge Army Camp Will Be Officially Started Tomorrow," *Santa Barbara News-Press*, September 14, 1941, 1–2; Bender, 222. Some documents put the contract award at $17,382,670.

4. "Huge Carryalls Scoop First Dirt on Camp Project," *Santa Barbara News-Press*, September 15, 1941, 1–2; "Camp Beginning Hailed by Talks as Dirt Thrown," *Santa Maria Daily Times*, September 15, 1941, 1–2.

5. "Camp Beginning Hailed by Talks as Dirt Thrown," *Santa Maria Daily Times*, September 15, 1941, 1–2.

6. Ibid.

7. Ibid.; Purkiss, 4–5; "Swirling Dust Clouds Once Again Retard Workmen at Camp Cooke," *Santa Barbara News-Press*, September 27, 1941, 1–2; "Much-Needed Water to Reach Building Area of Camp Soon," *Santa Barbara News-Press*, September 29, 1941, 1, 12; "Delay in Material, Lack of Water Holds Up Hiring on Camp Project," *Santa Barbara News-Press*, September 30, 1941, 1–2; "Water, Power Holding Camp Building Work," *Santa Maria Daily Times*, October 6, 1941, 1, 4.

8. "Much-Needed Water to Reach Building Area of Camp Soon," *Santa Barbara News-Press*, March 29, 1941, 1, 12; "Concrete Pouring Work Speeds Up Camp Construction Project," *Santa Barbara News-Press*, October 3, 1941, 1–2; "Extra Shift Ordered to Speed Work at Armored Division Camp," *Santa Barbara News-Press*, October 7, 1941, 1–2; "Work Roars Forward on Huge Army Project," *Santa Barbara News-Press*, October 16, 1941, 9, 14.

9. "Full-Swing Camp Construction Nears as Preliminary Work Ends," *Santa Barbara News-Press*, October 5, 1941, 1–2; "Electric Power Reaches Camp; Work Speed-Up Set," *Santa Barbara News-Press*, October 10, 1941, 1–2.

10. "Campsite Struck by Dust Storm," *Santa Barbara News-Press*, October 14, 1941, 1, 2; "Trains to Roll Over Camp Cooke Spur Railroad Within Few Days," *Santa Barbara News-Press*, November 16, 1941, 1–2; "Camp Work Hits One-Third Mark," *Santa Barbara News-Press*, November 23, 1941; 1–2; Purkiss, 5–6.

11. "County's Army Camp Named for Cavalry Officer," *Santa Barbara News-Press*, September 25, 1941, 1; Otis Young, *The West of Philip St. George Cooke, 1809–1895* (Glendale, CA: Arthur H. Clark, 1955), 21, 322–23, 353.

12. "Camp Cooke Work Near Full Stride," *Santa Barbara News-Press*, October 13, 1941, 1, 10; Ron Fink, "The Ocean Park—Surf Beach Docent Handbook" (Ocean Beach Commission, Santa Barbara County, 2001), n.p.

13. "Barrier Bridge to Get Steel Surface," *Cooke Clarion*, November 26, 1943, 1. "Disaster Losses Total Millions," *Lompoc Record*, January 27, 1969, 1, and images of the damaged bridge on pages 1 and 10. After the Air Force acquired Camp Cooke, it renamed the bridge Surf Bridge.

14. Headquarters Ninth Corps Area moved to Fort Douglas, Utah, on January 3 1942. Seven months later the Army reorganized and renamed its nine corps areas and their respective CASC reporting units. CASC 1908 became Service Command Unit 1908 of the Ninth Service Command, headquartered at Fort Douglas. Soldiers mockingly adopted the abbreviation SCU to mean the "Sick, Crippled, and Useless."

15. Purkiss, 6; "General Orders No. 87, Activation of Corps Area Service Command Unit," Headquarters Ninth Corps Area, Presidio of San Francisco, California, October 5, 1941, in the author's possession. "Captain in Cooke Gets Promotion," *Santa Maria Daily Times*, December 9, 1941, 4.

16. Purkiss, 6–7.

17. "Linemen Work at Camp Cooke," *Santa Barbara News-Press*, 30 September 1941, 4; "Camp Construction Tempo Stepped Up as Crew Grows," *Santa Barbara News-Press*, October 4, 1941, 1–2; "Campsite Struck by Dust Storm," *Santa Barbara News-Press*, October 14, 1941, 1, 2.

18. Purkiss, 36–37.
19. Ibid., 6, 7; "Office Moving to Camp Cooke," *Santa Barbara News-Press*, November 19, 1941, 1–2.
20. Purkiss, 7.
21. "Years of Service Behind Camp Cooke Staff Officers," *Santa Barbara News-Press*, February 23, 1942, 1, 10; Purkiss, 10.
22. "Work Started on $399,000 Camp Cooke Project: WPA and War Office Furnish Funds for Job," *Santa Barbara News-Press*, April 16, 1942, 1; "More Work for Camp Cooke," *Santa Maria Daily Times*, April 17, 1942, 1, 6; "W.P.A. Work Halted at Camp Cooke," *Los Angeles Times*, July 30, 1942, 9; contract summary, Records of the Chief Engineer, General Correspondence with District, 1941–1945, RG 77, Box 144, File no. 600.1, National Archives and Records Administration (hereafter cited as NARA), College Park, MD.
23. "Cooke Contract Confirmation Is Given," *Santa Maria Daily Time*, August 4, 1942, 2; Purkiss, 22–23.

## Chapter 2

1. "Light Shot Out During Black-Out by Civilian," *Lompoc Record*, December 12, 1942, 1; "Block Wardens Named for City Blackouts," *Lompoc Record*, December 19, 1941, 1, 5; "Col. Bres Gives Instructions to Camp Civilians," *Lompoc Record*, December 19, 1941, 1, 8; "Miguelito Road Blocked as City Guards Reservoir," *Lompoc Record*, December 19, 1941, 1.
2. "Many Japanese Arrested Here," *Santa Maria Daily Times*, December 8, 1941, 1–2; "F.B.I. Rounding Up Japanese," *Santa Maria Daily Times*, December 8, 1941, 6; "Japanese Aliens to Register at Police Station," *Lompoc Record*, December 12, 1942, 1, 8; John V. McReynolds, *Vanished: Lompoc's Japanese* (Lompoc, CA: Press Box Productions, 2010), 86–89. McReynolds delivers a fascinating account of how Lompoc was turned on its head by war hysteria, racial prejudice, and the personal greed of a few Caucasians who benefited from the removal of the Japanese. Of the approximate one-hundred Japanese families residing in Lompoc before the war, only two chose to return after VJ Day.
3. Adrian Gilbert, *The Encyclopedia of Warfare: From Earliest Times to the Present Day* (Guilford, CT: The Lyons Press, 2003), 256–59.
4. Shelby L. Stanton, *Order of Battle U.S. Army, World War II* (Novato, CA: Presidio Press, 1984), 54–55; Purkiss, 17, 31–32; Army Post Engineer, "Camp Cooke Building Index," May 15, 1955, 26, in the author's possession.
5. Stanton, 55.
6. *What's My Name?* "What's My Name?" March 13, 1942, 1; "Editors and Staff Celebrate Birthday of Camp Newspaper," *Camp Cooke Clarion*, March 12, 1943, 1, 4.
7. "Filipino Artist Designs Masthead," *Cooke Clarion*, May 21, 1943, 1; "Comic Creator: Pagsi-

lang Rey Isip," http://www.lambiek.net/artists/i/isip_pagsilang.htm (accessed February 6, 2013).
8. Purkiss, 18; "Furnishing of Guest House No. 2 Completed by Scandinavian Group," *The Clarion*, March 26, 1943, 1, 5; "Regulations Listed for Using Guest House Facilities," *Cooke Clarion*, July 16, 1943, 1; "Camp Cooke Building Index," 28.
9. "Regulations Listed for Using Guest House Facilities," *Cooke Clarion*, July 16, 1943, 1.
10. "Army Day Throng, 10,000 Strong, Inspects Huge Military Center," *Santa Barbara News-Press*, April 7, 1942, 1–2.
11. Purkiss, 62–63; "Camp Cooke Building Index," 67, 68.
12. Donald DeNevi, *The West Coast Goes to War, 1941–1942* (Missoula, MT: Pictorial Histories Publishing Company, 1998), 33–34; Wesley F. Craven and James L. Cate, eds. *The Army Air Forces in World War II*, 7 vols. (Chicago: The University of Chicago Press, 1948–1958, Reprint Washington, D.C.: Office of Air Force History, 1983), Vol. 1, *Plans and Early Operations, January 1939 to August 1942*, 282–83. Another twist on the "diversion" theme was proposed by the commander of the California military Guard, Gen. Donovan. Donovan said the shelling was a ruse to distract attention away from the real purpose of the submarine's mission—that being to put ashore an "emissary" to contact someone and possibly take that person or persons away with them. See "Ellwood Shelling Termed 'Ruse,'" *Santa Barbara News-Press*, March 25, 1942, 1.
13. Craven and Cates, 283–85.
14. National Park Service, "Radar Station B-71 Trinidad Radar Station California," in *V for Victory: America's Home Front During World War II*, by Stan Cohen, ed. (Missoula, MT: Pictorial Histories Publishing, 1991), 208; letter from Emil O. Lindner to Martin Hagopian, Deputy Historian, Vandenberg AFB, November 3, 1979, in the author's possession. The IV Interceptor Command was re-designated the IV Fighter Command in May 1942. The Coast Guard Rescue Station and Lookout Tower was established in December 1936. It was disestablished in 1952, but the property remained under their jurisdiction until February 1958, when it was acquired by the U.S. Navy. During the war years the facility was used more for coastal defense than as a rescue station. See David Gebhard and David Bricker, University of Santa Barbara, "The Former U.S. Coast Guard Lifeboat Rescue Station and Lookout Tower Point Arguello, California: (1936–1941), a Public Report," 1980.
15. Lindner to Hagopian.
16. Ibid.
17. Ibid.; Ava F. Kahn, "Historic Background of Red Roof Canyon," Social Process Research Institute, Public Archaeology, University of California, Santa Barbara, August 1981, 15.
18. "Fem Oil Checkers Hold Man-Sized Job," *Cooke Clarion*, February 11, 1944, 4. Former Ger-

man prisoner of war Adolf Kelmer also recalled seeing American women driving Army trucks at Camp Cooke. See Geiger, *German Prisoners of War at Camp Cooke, California: Personal Accounts, 1944–1946,* 83.

19. "Fem Oil Checkers Hold Man-Sized Job," *Cooke Clarion,* February 11, 1944, 4.

20. Ibid.

21. Purkiss, 40.

22. "Civilian Dormitories Being Constructed," *Cooke Clarion,* October 1, 1943, 1; "Camp Cooke Contract Let," *Santa Barbara News-Press,* September 23, 1943, B-6.

23. Wilson C. Von Kessler, "Annual Report of the Camp Surgeon, Camp Cooke, California, for the Year 1944," February 27, 1945, 4, Office of the Surgeon General, World War II Administrative Records, Unit Annual Reports, RG 112, Box 164, NARA, College Park, MD; "Cafeteria Ready for Civilian Use," *Cooke Clarion,* April 28, 1944, 1; "Civilian Cafeteria to Serve Evening Meals," *Cooke Clarion,* June 16, 1944, 6.

24. "Federal Trailer Camp to Open Here Wednesday," *Lompoc Record,* March 27, 1942, 1, 12.

25. "Formal Dedication Lompoc Gardens Set for Saturday," *Lompoc Record,* October 16, 1942, 1; "New 'Lompoc Terrace' Housing Open for Civilian Employees," *Cooke Clarion,* September 10, 1943, 1; "Camp Cooke Contract Let," *Santa Barbara News-Press,* September 23, 1943, B-6; advertisement from Cleveland Wrecking Company, *Lompoc Record,* April 29, 1954, II-8.

26. "Lompoc Will Acquire Hospital When Army Post Is Constructed," *Lompoc Record,* June 6, 1941, 1, 5; "Hospital Bid Accepted by FWA," *Lompoc Record,* August 2, 1946, 1.

27. Gilbert, 256, 260–63, 266–67.

28. "Men Now Over 38 Eligible for Discharge," *Camp Cooke Clarion,* February 12, 1943, 1; pamphlet, First and Second Filipino Infantry Regiments Association, *29th Anniversary and Fifth Annual Reunion,* July 30 through August 1, 1971, in author's possession; U.S. Army Center of Military History, "Asian Pacific Americans in the U.S. Army," http://www.history.army.mil/html/topics/apam/filipino_reg/filipino_regt.html (accessed April 2, 2012); Alex S. Fabros, "California's Filipino Infantry," reproduced at California State Military Museum, http://www.militarymuseum.org/Filipino.html (accessed April 2, 2012). The dates mentioned in these documents vary slightly. I chose the most plausible ones.

29. Stanton, 55, 56.

30. "Training at High Tempo Here During Last Year," *Cooke Clarion,* March 17, 1944, 7.

31. "Soldiers Gambol in Shady Groves of Oak Canyon's GI Playground," *Camp Cooke Clarion,* January 15, 1943, 4; "Training at High Tempo Here During Last Year," *Camp Cooke Clarion,* March 17, 1944, 7.

32. "Inoculation [*sic*] Course Realistic," *The Clarion,* May 14, 1943, 4; Robert R. Palmer, Bell I. Wiley, and William R. Keast, *Procurement and Training of Ground Combat Troops* (Washington, D.C.: Center of Military History, U.S. Army, 1991), 387.

33. "Inoculation [*sic*] Course Realistic," 4; Palmer et al., 387–88, 449, 452.

34. "Nazi Village Laid Out Here," *Cooke Clarion,* July 2, 1943, 3; Palmer et al., 387–88, 449.

35. "Nazi Village Laid Out Here," 3.

36. "6th Training in Night Fighting," *Cooke Clarion,* November 5, 1943, 1, 4; "Tanks Roll, Big Guns Roar as Sixth Armored Division Undergoes Intensive Tests," *Cooke Clarion,* July 30, 1943, 1–2.

37. "6th Armored Division Maintenance Battalion Demonstrates Efficiency," *Cooke Clarion,* October 29, 1943, 1, 4.

38. Ibid.

39. "6th Gets Swim'ng Training," *Cooke Clarion,* September 10, 1943, 1, 3; "Swimming Tests Qualify 1,000 Division Reconnaissance Men," *Cooke Clarion,* September 17, 1943, 5.

40. "Camp Cooke's Mechanized Might Shown in War Bond Parades," *Cooke Clarion,* September 17, 1943, 1, 3; "City Welcomes First Contingent of Camp Cooke Soldiers; Parade Due Thursday," *Santa Barbara News-Press,* September 7, 1943, 1, A-2; Tom Kleveland, "Mock Tank War Shows Need for Buying Bonds," *Santa Barbara News-Press,* September 12, 1943, 1, A-4.

41. "Stream-Crossing Expedients Demonstrated by 21st Infantry," *Cooke Clarion,* August 25, 1944, 1, 3; *Cooke Clarion,* photo caption, December 3, 1943, 6.

42. "Division—Special Troop Units in Giant Maneuver," *Cooke Clarion,* December 10, 1943, 1, 3.

43. "Commanding General 4th Army Inspects Troops," *Cooke Clarion,* December 3, 1943, 1, 3; Stanton, 66.

44. "Division—Special Troop Units in Giant Maneuver," *Cooke Clarion,* December 10, 1943, 1, 3.

45. Ibid.

46. "Sefor Torah Dedicated at Jewish Services," *Cooke Clarion,* June 4, 1943, 1; "Chapel Services Directory," *Cooke Clarion,* December 3, 1943, 2.

47. "Filipinos, Flashing Bolos, Will Stage Review Saturday," *Cooke Clarion,* July 16, 1943, 1; "... Fight with Bolos," *Cooke Clarion,* July 23, 1943, 1, 3; "Filipinos Train to Fight with Bolo Knives, *Los Angeles Times,* August 1, 1943, 22.

48. "Filipino Radio Records Made," *Cooke Clarion,* September 24, 1943, 1, 6; "Filipino, Chinese Soldiers Make OWI Overseas Recordings," *Cooke Clarion,* April 28, 1944, 1, 6.

49. "Cooke Bank Opens Monday July 26, 1000," *Cooke Clarion,* July 28, 1943, 1; "New Camp Cooke Branch of the Bank of America Opens Doors for Business," *Cooke Clarion,* July 30, 1943, 1.

50. "Col. Brumbaugh New CO," *Cooke Clarion,* September 24, 1943, 1, 3.

51. "Santa Maria Harvest Emergency Will Be Relieved by Soldier Aid," *Cooke Clarion*, 1 October 1943, 1; "Top Soldier Tomato Picker Makes $13 with Crop Peak Yet to Come," *Cooke Clarion*, October 15, 1943, 1, 3; "Tomato Harvest Emergency Over," *Cooke Clarion*, November 12, 1943, 1.

52. Purkiss, 44; "WAC Officer First Here," *Cooke Clarion*, October 1, 1943, 1.

53. Purkiss, 44; "Last Enlisted WAC Goes to New Station," *Cooke Clarion*, April 13, 1945, 1; "Lt. Miller, WAC, Departs Wearing Engineer Insigne," *Cooke Clarion*, April 13, 1945, 1.

54. Purkiss, 44, 45.

55. "Increase Is Seen for Next Report as Interest Grows," *Camp Cooke Clarion*, June 12, 1942, 1, 5; "Battle Royal Looms in Camp Bond Drive Among Civilian Units," *Camp Cooke Clarion*, November 13, 1942, 1; "Treasury 'T' Flag for War Bond Purchases Earned by Camp Cooke," *Cooke Clarion*, February 18, 1944, 1, 3; "100% Goal Sought in Fifth War Loan Drive," *Cooke Clarion*, June 2, 1944, 1, 4.

56. "Camp Assists War Bond Rally at High School," *Cooke Clarion*, May 28, 1943, 1–2.

57. Ibid.

58. Ibid.

59. Event advertisement, *Santa Maria Daily Times*, August 3, 1943, 6; "Cooke Units Assist Santa Maria in War Bond Fiesta," *Cooke Clarion*, August 6, 1943, 1–2; "Miss Gerlich Is Bond Queen," *Santa Maria Daily Times*, August 5, 1942, 1–2; "City Gives Over Day to Boosting Bond Sales," *Santa Maria Daily Times*, August 4, 1942, 1–2.

60. "Cooke Units Assist Santa Maria in War Bond Fiesta," *Cooke Clarion*, August 6, 1943, 1–2; "City Gives Over Day to Boosting Bond Sales," *Santa Maria Daily Times*, August 4, 1942, 1–2.

61. "Cooke Units Assist Santa Maria in War Bond Fiesta," *Cooke Clarion*, August 6, 1943, 1–2; "Bond Auction Brings $150,000 to Local Drive," *Santa Maria Daily Times*, August 5, 1943, 1–2.

62. "Miss Gerlich Is Bond Queen," *Santa Maria Daily Times*, August 5, 1942, 1–2; "Cooke Units Assist Santa Maria in War Bond Fiesta," *Cooke Clarion*, August 6, 1943, 1–2; *Cooke Clarion*, photo caption, August 13, 1943, 3. The caption incorrectly lists the event as having occurred on July 28, 1943.

63. Gilbert, 261.

64. Margaret E. Wagner, Linda B. Osborn, Susan Reyburn, and the Staff of the Library of Congress, eds., *World War II Companion* (New York: Simon & Schuster, 2007), 585–87, 603.

65. Ibid., 580–82.

66. Gilbert, 268–69.

67. Stanton, 63; "Brig. Gen. Kilburn Takes Command of 11th Division," *Cooke Clarion*, March 24, 1944, 1, 3; "Gen. Kilburn, in First Talk to 11th, Announces 'Thunderbolt,'" *Cooke Clarion*, April 7, 1944, 1, 4.

68. "Filipino Awarded Soldier's Medal," *Cooke Clarion*, March 31, 1944, 1, 3; "P'master General Visitor," *Cooke Clarion*, March 31, 1944, 1.

69. "General Marshall Visits Here," *Cooke Clarion*, May 19, 1944, 1, 3.

70. House Committee on Claims, *Vertie Bea Loggins*, 79th Cong., 2nd sess., 1946. H. Rep. 1762, 1–5; "Shell Fragments at Camp Cooke Hit 'Daylight,'" *Santa Maria Daily Times*, May 12, 1944, 1–2.

71. House Committee on Claims, *Vertie Bea Loggins*, 79th Cong., 2nd sess., 1946. H. Rep. 1762, 1–5.

72. Ibid.; Willedd Andrews to Representative Clyde Doyle, April 11, 1945, Helen Gahagan Douglas Collection (hereafter cited as HGDC), University of Oklahoma; Willedd Andrews to Representative Helen Gahagan Douglas, September 29, 1945, HGDC; Representative Helen Gahagan Douglas to Willedd Andrews, October 23, 1945, HGDC; Representative Helen Gahagan Douglas to House Claims Committee Clerk, October 26, 1945, HGDC; Willedd Andrews to Representative Helen Gahagan Douglas, October 29, 1945, HGDC; Representative Helen Gahagan Douglas to Willedd Andrews, June 14, 1946, HGDC; Daniel Mark Epstein, *Sister Aimee: The Life of Aimee Semple McPherson* (New York: Harcourt Brace Jovanovich, 1993), 403.

73. Willedd Andrews to Representative Helen Gahagan Douglas, October 7, 1945, HGDC.

74. "Ab Jenkins, Famed Speed King, at Cooke as Maintenance Advice," *Cooke Clarion*, May 19, 1944, 1, 3.

75. Jeffrey E. Geiger, *German Prisoners of War at Camp Cooke, California: Personal Accounts of 14 Soldiers, 1944–1946* (Jefferson, NC: McFarland, 1996), 1.

76. Ibid., 1–2.

77. Ibid., 69, 70.

78. "Work of Italian Service Units Here Described," *Cooke Clarion*, April 6, 1945, 1, 3.

79. "142nd Italian Service Unit Leaves Monday," *Cooke Clarion*, June 22, 1945, 1; "3rd Italian Service Unit Arrives in Camp," *Cooke Clarion*, August 3, 1945, 1; "ISU Outfits Leave for New Station," *Cooke Clarion*, September 7, 1945, 3.

80. "1000 Division Catholic Soldiers Attending Religious Retreat," *Cooke Clarion*, June 23, 1944, 1, 4; "Catholic Soldiers Pray at Purisima," *Cooke Clarion*, June 30, 1944, 1, 4.

81. "Navy Cooperates with Division Problem," *Cooke Clarion*, August 18, 1944, 4.

82. "Camp Cooke's Airport 'Just Grew,'" *Cooke Clarion*, January 14, 1944, 3; "Cooke Landing Strip Lengthened 600 Ft.," *Cooke Clarion*, May 12, 1944, 1. The current runway at Vandenberg AFB was built over the Army's runway.

83. "War Dogs Detect Mines, Deliver Messages in Division Exhibition," *Cooke Clarion*, August 4, 1944, 6.

84. "'Mustard Gas' Adds to Cooke's Realistic Combat Training Course," *Cooke Clarion*, August 11, 1944, 1, 3.

85. "Division Marks 2nd Birthday," *Cooke Clarion*, August 18, 1944, 1, 3.

86. Stanton, 158.

87. Ibid., 173.

88. "Creeping, Crawling, Using All Typical Infantry Weapons, 386th Troops Demolish 'Enemy' Pillbox in Training Problem," *Cooke Clarion*, October 27, 1944, 1.

89. "Religious Event Draws 6000," *Cooke Clarion*, November 24, 1944, 1, 4. In 1954 Rabbi Nussbaum took the pulpit at Beth Israel in Jackson, Mississippi, and became an activist in the early Civil Rights movement. After local Ku Klux Klan members bombed his synagogue and residence in 1967, he also became chaplain to the jailed Freedom Riders and surreptitiously relayed messages to them from family members in the North because authorities refused them contact. Nussbaum moved to San Diego, California, in 1973. He was 79 when he died in 1987.

90. Sgt. Frank Valenti, "Twenty-Five Hundred Catholic Soldiers Take Part in Impressive 2-Day La Purisima Retreat," *Cooke Clarion*, December 1, 1944, 1, 4.

91. Wagner, ed., et al., 599, 604; Congressional Medal of Honor Society, http://www/cmohs.org/recipient-detail/2750/gammon-archer-t.php (accessed February 14, 2012).

92. Wagner, ed., et al., 611, 617–19.

93. Ibid., 608, 613.

94. "Personnel Attend Memorial Service," *Cooke Clarion*, April 20, 1945, 1, 3; "Westmont Choir Sings at Hospital," *Cooke Clarion*, April 20, 1945, 1.

95. "Westmont Choir Sings at Hospital," *Cooke Clarion*, April 20, 1945, 1.

96. Gilbert, 273; "20th Named as Second Armored Outfit Due," *Cooke Clarion*, July 20, 1945, 1; "Advance Unit of 13th AD Arrives with Records," *Cooke Clarion*, August 3, 1945, 1; "20th Armd. Div. Commanding General and Advance Party Arrive," *Cooke Clarion*, September 7, 1945, 1; Stanton, 66, 69; "764th Water Buffaloes Arrive for Training," *Cooke Clarion*, July 20, 1945, 1.

97. "Preparations Speeded as Heavy Equipment Arrives," *Cooke Clarion*, July 20, 1945, 1; "Material Continues Rolling In," *Cooke Clarion*, August 17, 1945, 3.

98. "Training Areas Being Built Here," *Cooke Clarion*, August 3, 1945, 1.

99. Ibid.

100. "First of Point Men Leave as 38-Yr. Olds Get Set," *Cooke Clarion*, August 24, 1945, 1, 3; "All Men Eligible for Discharge 'Turned Loose,'" *Cooke Clarion*, September 28, 1945, 1; Wesley F. Craven, and James L. Cate, eds., *The Army Air Forces in World War II*. 7 vols. (Chicago: The University of Chicago Press, 1948–1958; reprint, Washington, D.C.: Office of Air Force History, 1983), Vol. 7, *Service Around the World*, note, 568.

101. "Separation Point Begin Operations on Saturday," *Cooke Clarion*, September 14, 1945, 1–2; "Separation Point Closes Up Shop," *Cooke Clarion*, December 14, 1945, 1, 3.

102. "Thousands Take Separation Route," *Cooke Clarion*, January 25, 1946, 1; "Schedule of Separation Center-Bound Trains for 20th Armd. Div., VII Corps Troops Listed," *Cooke Clarion*, February 1, 1946, 1.

103. "13th Armd. Div. Winds Up U.S. Army Career," *Cooke Clarion*, November 16, 1945, 1–2; "Storage of 13th Armd. Div. Vehicles Is More Than Just 'Leaving 'em Be,'" *Cooke Clarion*, November 16, 1945, 6.

104. "All AGF Units Here Either Inactivating or Leaving; Camp Closing Expected in March," *Cooke Clarion*, February 8, 1946, 1; "War Department Orders Camp Cooke Closing," *Lompoc Record*, February 7, 1946, 1; "20th Division Leaves for Texas Monday," *Cooke Clarion*, March 1, 1946, 1; Stanton, 69.

105. "Guest House 1 Closes; No. 2 Remains Open," *Cooke Clarion*, January 18, 1946, 1; "Guest House to Close," *Cooke Clarion*, March 1, 1946, 1; "Service Club to Close," *Cooke Clarion*, March 1, 1946, 1; "Theaters 1, 2 to Close," *Cooke Clarion*, March 1, 1946, 1; "4 PXs Open," *Cooke Clarion*, March 1, 1946, 1; "Chapels C, D Close," *Cooke Clarion*, March 8, 1946, 1; "Branch Bank Closes March 29," *Cooke Clarion*, March 15, 1946, 1; "Post Red Cross Sets March 31 Closing Date," *Cooke Clarion*, March 22, 1946, 1; "Clarion Quits After Four Years," *Cooke Clarion*, April 5, 1946, 1, 3.

106. "Camp Cooke on 'Standby' Basis Starting June 1," *Santa Maria Daily Times*, March 15, 1946, 1; "Temporary Buildings in Camp Cooke to Be Sold as Surplus," *Santa Maria Daily Times*, July 16, 1946, 1.

## *Chapter 3*

1. "Contractor Breaks Ground for $4,500,000 Project," *Lompoc Record*, July 6, 1945, 1; "Annual Report AG 10, 1 July 1948 to 30 June 1949," Disciplinary Barracks, Records of Headquarters Army Ground Forces, RG 337, Box 18, NARA, College Park, MD.

2. "Contractor Breaks Ground for $4,500,000 Project," *Lompoc Record*, July 6, 1945, 1; "1,500 to be Housed in Military Prison," *Santa Maria Daily Times*, September 24, 1945, 1–2; "Annual Report AG 10, 1 July 1948 to 30 June 1949," Disciplinary Barracks, Records of Headquarters Army Ground Forces, RG 337, Box 18, NARA, College Park, MD; Bender, 266.

3. "Annual Report AG 10, 1 July 1948 to 30 June 1949," Disciplinary Barracks, Records of Headquarters Army Ground Forces, RG 337, Box 18, NARA, College Park, MD; Bender, 266.

4. "Disciplinary Units Arriving at Cooke," *Santa Maria Daily Times*, November 7, 1946, 1; "Annual Report AG 10, 1 July 1948 to 30 June

1949," Disciplinary Barracks, Records of Headquarters Army Ground Forces, RG 337, Box 18, NARA, College Park, MD; "Disciplinary Barracks to Be Activated December First," *Lompoc Record*, November 7, 1946, 1, 10; "USDB Activation to Be Immediate," *Lompoc Record*, December. 12, 1946, 1.

5. "Annual Report AG 10, 1 July 1948 to 30 June 1949," Disciplinary Barracks, Records of Headquarters Army Ground Forces, RG 337, Box 18, NARA, College Park, MD; *Lompoc Record*, "D. B. Activated with Arrival of Prisoners," January 23, 1947, 1.

6. "Parole Officer Tells Rotary About Prison at Camp Cooke," *Santa Maria Daily Times*, September 17, 1947, 6.

7. Advertisement, "Buildings for Sale," Bosley Wrecking Company, *Santa Maria Daily Times*, November 28, 1947, 9; "Annual Report AG 10, 1 July 1948 to 30 June 1949," Disciplinary Barracks, Records of Headquarters Army Ground Forces, RG 337, Box 18, NARA, College Park, MD.

8. Geiger, *German Prisoners of War at Camp Cooke*, 147–48; "14 POW Bodies Being Sent to San Bruno," *Santa Maria Daily Times*, November 25, 1947, 2.

9. "Most of Camp to Be Leased for 5 Years," *Lompoc Record*, December 9, 1948, 1; "Initial OK on Cooke Leases," *Lompoc Record*, January 6, 1949, 1; "More Cooke Acreage Offered for Lease," *Lompoc Record*, January 20, 1949, 1; "Cooke Cantonment Area Offered for Sheep Grazing," *Lompoc Record*, June 2, 1949, 1; "Vast Acreage at Cooke Offered for Civilian Use," *Lompoc Record*, November 17, 1949, 1.

10. "Secy. of Defense, Wedemeyer View Guard at Cooke," *Lompoc Record*, July 6, 1950, 1; "Nevada Governor Visits Guard Unit at Cooke," *Lompoc Record*, July 27, 1950, II-5.

## Chapter 4

1. Roy E. Appleman, *South to the Naktong, North to the Yalu (June–November 1950)* (Washington, D.C.: Center of Military History, U.S. Army, 1961), 59, 68, 74–75.

2. Ibid., 37–39.

3. Gilbert, 280–81.

4. Ibid., 281.

5. "Headquarters Camp Cooke, California: Official Diary," n.d., 3 (hereafter cited as "Official Diary"), AGF and AFF Installations, Camp Cooke, Records of Headquarters Army Ground Forces, RG 337, Box 17, NARA, College Park, MD; "Call 40th Into Federal Service," *Santa Maria Times*, August 1, 1950, 1, 3; "DB, Camp Cooke 'Divorced' by Army Action," *Santa Maria Times*, August 16, 1950, 1, 8; program booklet, Army Special Services, *Camp Cooke Welfare Carnival*, October 1952, n.p., Lompoc Valley Historical Society.

6. "Official Diary," 55; "Cancel Leases on Camp Cooke Range Lands," *Lompoc Record*, August 31, 1950, 1; "Cooke Leases Move Cattle from Camp," *Santa Maria Times*, September 18, 1950, 6.

7. "Open Bids for Cooke Roads," *Santa Maria Times*, September 2, 1950, 8.

8. "Official Diary," 11.

9. "Guard Division Moves Into Camp Cooke for Training," *Santa Maria Times*, September 6, 1950, 1; "Draftees Arrive to Fill in 40th," *Santa Maria Times*, October 7, 1950, 5; "Film Actor Tom Brown Made Captain," *Los Angeles Times*, July 5, 1950, A3; "Lompocans to Be Featured on 40th Division Show," *Lompoc Record*, September 28, 1950, II-2; Congressional Medal of Honor Society, http://www.cmohs.org (accessed February 14, 2012); "84 Men from Guam Arrive at Camp Cooke," *Santa Maria Times*, November 28, 1950, 1; "On the Sound Track," *The Billboard*, September 2, 1950, 21; Hedda Hopper, "'No Questions Asked' Stars Dahl, Sullivan," *Los Angeles Times*, October 3, 1950, A10. Lilley would serve with distinction in Korea, earning two Bronze Stars and a promotion to Major. See "Major Lilley Receives 2nd Bronze Star," *Lompoc Record*, August 7, 1952, I-7.

10. "First Detachment of WACs Arrive at Camp Cooke," *Lompoc Record*, September 21, 1950, 5; "Official Diary," 8; "Camp Cooke WAC Unit Gets New Commanding Officer," *Santa Maria Times*, November 20, 1950, 12.

11. "Cooke MPs to Observe Birth of Their Corps," *Santa Maria Times*, September 27, 1951, 5; program booklet, *Camp Cooke Welfare Carnival*, October 1952, n.p.

12. "Camp Cooke 'Clarion' Makes First Appearance Since '46," *Santa Maria Times*, September 29, 1950, 1; see services section of *Cooke Clarion*, February 23, 1951, 2; "Official Diary," 3, 4, 12, 13.

13. "Part of Cooke Hospital Back in Use by Army," *Lompoc Record*, April 29, 1948, 1; "Annual Report of Post Surgeon," Camp Cooke, CA, to the U.S. Surgeon General, February 5, 1951, 2, AGF and AFF Installations, Camp Cooke, Records of Headquarters Army Ground Forces, RG 337, Box 16, NARA, College Park, MD.

14. "Station Hospital Expanding Rapidly," *Lompoc Record*, September 14, 1950, I-7; "Annual Report of Medical Activities, 1952," Camp Cooke, CA, to the U.S. Surgeon General, n.d., n.p., AGF and AFF Installations, Camp Cooke, Records of Headquarters Army Ground Forces, RG 337, Box 16, NARA, College Park, MD; "Annual Report of Army Medical Services Activities, 1951," Camp Cooke, CA, to the U.S. Surgeon General, January 15, 1952, 3, 6, 27, 47, AGF and AFF Installations, Camp Cooke, Records of Headquarters Army Ground Forces, RG 337, Box 16, NARA, College Park, MD.

15. "Annual Report of Post Surgeon," Camp Cooke, CA, to the U.S. Surgeon General, February 5, 1951, 11, AGF and AFF Installations, Camp Cooke, Records of Headquarters Army Ground Forces, RG 337, Box 16, NARA, College Park, MD; "Annual Report of Army Medical Services Activities, 1951," Camp Cooke, CA, to the U.S. Surgeon General, January 15, 1952, 25, AGF and AFF Installations, Camp Cooke, Records of Headquarters Army Ground Forces, RG 337, Box 16, NARA, College Park, MD; "Annual Report of Medical Activities, 1952," Camp Cooke, CA, to the U.S. Surgeon General, n.d., 4, 12, AGF and AFF Installations, Camp Cooke, Records of Headquarters Army Ground Forces, RG 337, Box 16, NARA, College Park, MD; "Bullet Takes 44th Soldier," *Santa Maria Times*, April 1, 1952, 1.

16. "Official Diary," 61; "Annual Report of Army Medical Services Activities, 1951," Camp Cooke, CA, to the U.S. Surgeon General, January 15, 1952, 94–95, AGF and AFF Installations, Camp Cooke, Records of Headquarters Army Ground Forces, RG 337, Box 16, NARA, College Park, MD.

17. "Annual Report of Army Medical Services Activities, 1951," Camp Cooke, CA, to the U.S. Surgeon General, January 15, 1952, 98–99, AGF and AFF Installations, Camp Cooke, Records of Headquarters Army Ground Forces, RG 337, Box 16, NARA, College Park, MD; "Local Organizations Doing Their Bit for Veterans in Camp Cooke Hospital," *Lompoc Record*, September 25, 1952, 3; "Official Diary," 68.

18. "Colonel Williams Reveals Army Housing Plight," *Lompoc Record*, October 26, 1950, 1.

19. "Governor Warren Pays Visit to 40th Infantry Division," *Santa Maria Times*, October 24, 1950, 1.

20. "40th Plays Host to Businessmen," *Santa Maria Times*, October 27, 1950, 1; "40th Division to Demonstrate Arms Firing," *Santa Maria Times*, November 28, 1950, 1.

21. *747th Amphibious Tank and Tractor Battalion, Camp Cooke, California* (Baton Rouge: Army and Navy Publishing, *circa* 1952), n.p. This book is mostly a pictorial history of the armored vehicles and battalion personnel.

22. E-mail from Philip Videon to the author, February 10, 2010.

23. Ibid., October 16, 2012.

24. *747th Amphibious Tank and Tractor Battalion*, n.p.; "Amphibs Get Underway for Coronado Maneuver," *Cooke Clarion*, Aug 17, 1951, 1; miscellaneous newspaper clipping, "Amphibious Power Shown on Strand," August 1951, provided to the author by Philip Videon.

25. "Official Diary," 37–8; "Rear Detachment Remains to Train 4000 Replacement-Troops for 40th," *Cooke Clarion*, March 21, 1951, 5; Walter G. Hermes, *Truce Tent and Fighting Front* (Washington, D.C.: Center of Military History, U.S. Army, 1966), 202–03.

26. "Headquarters, 3265th Training Squadron, United States Air Force Historical Research Agency, Maxwell Air Force Base, Montgomery, AL, microfilm roll K0707, index 1944, 1092–1139; "Air Forcers at Cooke Say 'Bye to Army," *Santa Maria Times*, September 3, 1952, 1. This article incorrectly lists the squadron number as the 3256th and wrongly states that it arrived at Cooke in February 1951.

27. "Official Diary," 65.

28. "Lompocans See Atom Blast Light Up Sky Monday," *Lompoc Record*, February 8, 1951, I-1, 8; "Soldiers Observe Atomic Blast During Reveille," *Lompoc Record*, February 8, 1951, II-7; M. Barrett et al., *Analysis of Radiation Exposure for Military Participants: Exercises Desert Rock I, II, and III—Operation Buster-Jangle*, Science Applications International Corp., December 1987, under contract to U.S. Defense Nuclear Agency, 1, 8, http://www.dtra.mil/documents.ntpr/relatedpub/DNATR87 116.PDF (accessed February 4, 2013).

29. Barrett et al., 10–11, 22–23, 32; Jean Ponton et al., *Shots Sugar and Uncle: The Final Tests of the Buster-Jangle Series, 19 November–29 November 1951*, JRB Associates, June 23, 1982, under contract to the U.S. Defense Nuclear Agency, 13, 21–22, 27, http://www.dtra.mil/documents.ntpr/historical/1951 (accessed February 4, 2013).

30. Ponton et al., 13–14, 24, 27.

31. "Cooke Soldiers at Atomic Test Safe from Hurt," *Santa Maria Times*, September 21, 1951, 1.

32. "Official Diary," 92. This report mentions only one soldier, Pvt. Kenneth W. Bell, and gave his arrival date as March 18. An Army hospital report prepared at Cooke indicates the hospital received its first two battle casualties on March 19 but omits their names. See "Hospital Daily Diary, 8 January 1952–31 December 1952," n.p., AGF and AFF Installations, Camp Cooke, Records of Headquarters Army Ground Forces, RG 337, Box 17, NARA, College Park, MD; "Annual Report of Army Medical Services Activities, 1951," Camp Cooke, CA, to the U.S. Surgeon General, January 15, 1952, 26, AGF and AFF Installations, Camp Cooke, Records of Headquarters Army Ground Forces, RG 337, Box 16, NARA, College Park, MD.; "Annual report of Medical Activities for the Year 1952, 2, AGF and AFF Installations, Camp Cooke, Records of Headquarters Army Ground Forces, RG 337, Box 16, NARA, College Park, MD.

33. "Camp Cooke Pulls All Stops Out for Armed Forces Day 'Special,'" *Santa Maria Times*, May 18, 1951, 1, 6; "7,000 Spectators Attend Armed Forces Day at Camp Cooke to See 'Fire Power,'" *Santa Maria Times*, May 21, 1951, 1; "Armed Forces Day Draws Thousands to Camp Cooke," *Lompoc Record*, May 24, 1951, II-3; "First Korean Evacuee Dies at Cooke," *Santa Maria Times*, April 28, 1951, 1; "Port Hueneme Officer to Talk Here on Armed Forces Program," *Santa Maria Times*, May 7, 1951, 1.

34. "7,000 Spectators Attend Armed Forces Day at Camp Cooke to See 'Fire Power,'" *Santa Maria Times*, May 21, 1951, 1.

35. "Governor Warren Observes National Guard Training," *Santa Maria Times*, June 29, 1951, 3.

36. "Official Diary," 46, 49, 51; "Cooke Greets NG and ORC," *Cooke Clarion*, June 15, 1951, 1;"Metallic Monsters in Mass Moves as 13th Armored Roars into Action," *Cooke Clarion*, August 10, 1951, 1; "Completed ORC Program Vital to Defense Plans," *Cooke Clarion*, August 17, 1951, 1; "49th Division to Profit from Korean War Tactics," *Santa Maria Times*, June 18, 1951, 1, 8; "Reserve National Guard Divisions at Camp Cooke," *Lompoc Record*, June 28, 1951, I-7.

37. "Divot Diggers to Tee-Off Sunday as Camp Cooke Golf Course Opens," *Cooke Clarion*, July 27, 1945, 1; "Teeing Off," *Cooke Clarion*, July 27, 1951, 4.

38. "400 Soldiers to Help Pick Tomato Crop," *Santa Maria Times*, October 20, 1951, 1.

39. "Official Diary," 65; "Colonel Kirby Is Top Man at Cooke," *Santa Maria Times*, November 21, 1951, 1.

40. "Unit History, 466th AAA BN," circa May 1952, and "Historical Data Concerning Antiaircraft Defense," n.d., Unit Histories, 1940–1970, Records of U.S. Army Operational, Tactical, and Support Organizations (World War II and Thereafter), RG 338, Box 1292, NARA, College Park, MD; "466th Battalion Makes Self at Home at Camp Cooke; Completing Training," *Lompoc Record*, January 24, 1952, 1.

41. "Official Diary," 82, 87, 90; "44th Material Is Pouring In," *Santa Maria Times*, February 13, 1952, 1; "44th Rolls in at Camp Cooke," *Santa Maria Times*, February 21, 1952, 1, 6; "Homeless Families of 44th Trekking All Over Vicinity," *Santa Maria Times*, February 19, 1952, 1, 3.

42. Carroll Gewin, "44th Will Be Called on for Troops to Fight Korean War, Clark Says," *Santa Maria Times*, April 24, 1952, 1; "2000 Men of 44th Ordered Overseas," *Lompoc Record*, September 4, 1952, II-1; "Korea Returnees to Bolster 44th," *Santa Maria Times*, October 13, 1952, 1.

43. "Stevenson Visits Cooke," *Santa Maria Times*, May 7, 1952, 1; "Illinois Governor Witnesses First Full-Dress Review by 44th Division," *Santa Maria Times*, May 8, 1952, 1.

44. "Official Diary," 107–08; "'Mad Minute' to Feature Camp Cooke Open House," *Santa Barbara News-Press*, May 16, 1952, B-1; "Public Invited to Armed Forces Day at Cooke," *Santa Barbara News-Press*, May 11, 1952, A-4.

45. "747th Starts Program to Raise Educational Level of Battalion," *Cooke Clarion*, March 21, 1951, 5.

46. "Official Diary," 140.

47. Ibid.

48. Ibid.

49. "To Begin Extension Courses Here, 3 Apr.," *Cooke Clarion*, March 21, 1951, 1.

50. "Official Diary," 129–30.

51. U.S. Department of State, Office of the Historian, "Milestones: 1942–1952," http://www.history.state.gov/milestones/1945-1952/NSC68 (accessed November 13, 2012), 1; "'Watchdog' Prober Raps Continued Military Waste," *Los Angeles Times*, September 15, 1952, 8; "Eisenhower Offers Defense Solvency," *Los Angeles Times*, September 26, 1952, 1, 6; brochure, "Welcome to Camp Cooke," U.S. Army, 1952, in the author's possession.

## Chapter 5

1. "First USO Dance To Be Given Friday Evening," *Lompoc Record,* January 30, 1942, 4; "Servicemen to be Entertained by Townspeople," *Lompoc Record,* February 13, 1942, 2.

2. "Peak Special Service Facilities," *Cooke Clarion*, April 5, 1946, 3.

3. Purkiss, 18, 49.

4. "Camp Cooke Service Club Center of Many Recreational Programs," *Camp Cooke Clarion*, June 12, 1942, 4; "Bob Hope Stages Broadcast in Camp Cooke," *Santa Maria Daily Times*, March 4, 1942, 2.

5. Bob Hope script, March 3, 1942, Milton Josefberg papers, collection 3154, box 7, American Heritage Center, University of Wyoming.

6. "Camp Cooke Service Club Center of Many Recreational Programs," *Camp Cooke Clarion*, June 12, 1942, 4.

7. "All Girl Revue and Harpo Marx Monday," *What's My Name?* March 13, 1942, 1–2, 6; United Service Organizations, "Camp Shows Publicity Records, 1941–1957," microfilm reel No. 1, 695, Margaret Herrick Library, Academy of Motion Picture Arts and Sciences, Beverly Hills, CA.

8. "Camp Cooke Service Club Center of Many Recreational Programs," *Camp Cooke Clarion*, June 12, 1942, 4.

9. "Peak Special Service Facilities," *Cooke Clarion*, April 5, 1946, 3; Army Post Engineer, "Camp Cooke Building Index," 27.

10. "Peak Special Service Facilities," *Cooke Clarion*, April 5, 1946, 3.

11. Ibid.

12. Ibid.; Purkiss, 18; Army Post Engineer, "Camp Cooke Building Index," 43.

13. "Dance Party Opens C.A.S.C. Officers Club," *Camp Cooke Clarion*, May 29, 1942, 1–2; Army Post Engineer, "Camp Cooke Building Index," 26.

14. "'3 Stooges' at Arena Next Week," *Camp Cooke Clarion*, May 22, 1942, 1; "Stooges Give Up $8,500 in Wages to Appear Here," *Camp Cooke Clarion*, May 29, 1942, 1, 6.

15. "'What's Cookin' Sports Arena Hit," *Camp Cooke Clarion*, June 12, 1942, 1, 5; "Radio Lost

Song Writer in Tepper to Talent Staff of 'V' Division," *Camp Cooke Clarion*, June 12, 1942, 1; Wikipedia, "Sid Tepper," http://en.wikipedia.org/wiki/Sid_Tepper (accessed January 15, 2013).

16. "'Room Service' Is Launch Riot," *Camp Cooke Clarion*, June 19, 1942, 1, 4.

17. "New Theater Ends Waiting for Show Seats," *Camp Cooke Clarion*, June 19, 1942, 1; "New Theater No. 3 to Open Tonight," *Camp Cooke Clarion*, February 26, 1943, 1.

18. "'Roxy Revue' at Arena Next Week," *Camp Cooke Clarion*, June 19, 1942, 1, 3.

19. "Eastern Tank Soldiers Get Taste of 'Rodey-o,'" *Los Angeles Times*, June 29, 1942, A1.

20. "USO to Observe Third Milestone in War Service," *Lompoc Record*, February 4, 1944, 1–2.

21. Meghan K. Winchell, *Good Girls, Good Food, Good Fun* (Chapel Hill: University of North Carolina Press, 2008), 18–19.

22. Purkiss, 18; "Officers to Speak at Dedication of Santa Maria's U.S.O. Building," *Santa Barbara News-Press*, March 15, 1942, 14; *Santa Maria Daily Times*, "USO Activities Planned for Thanksgiving Day," November 16, 1945, 1–2. The library was built in 1909 using a $10,000 grant from the Andrew Carnegie Foundation. It closed in June 1941 when a new library opened next door at 420 South Broadway. After World War II, the building served as a youth center under the city's Parks and Recreation Department, except for the period during the Korean War when it again opened as a USO clubhouse. In 1969 the building was torn down. See "History of the Santa Maria Public Library" on file in the library archives, updated June 16, 2009.

23. "U.S.O. Dormitory Opens Friday in Santa Maria," *Santa Barbara News-Press*, February 5, 1943, 5; "Open House to Mark U.S.O. Sunday in S.M.," *Santa Barbara News-Press*, March 28, 1943, 16.

24. Winchell, 8–9; "Pvt. Teddy Roosevelt Winston, Colored Singer-Dancer, 'Steals' Club 1 Variety Show," *Cooke Clarion*, August 11, 1944, 3; "USO Four Years Old This Week," *Santa Maria Daily Times*, January 30, 1945, 3; "USO Activities Planned for Thanksgiving Day," *Santa Maria Daily Times*, November 16, 1945, 1–2. Pvt. Winston was stationed at the Santa Maria Army Air Base.

25. "New Lompoc USO 'Hit' with Men," *Camp Cooke Clarion*, July 17, 1942, 6; "Hundreds Attend Opening of Foresters Hall as USO Soldier Recreation Center," *Lompoc Record*, July 17, 1942, 1, 8; USO brochure "'H' Street Club," n.d., Lompoc Valley Historical Society. Frank Dinwiddie built the original structure in 1876, from which he sold goods and lived with his family on the ground floor, and rented out the top floor. He later sold the building to the Foresters Lodge. Building history file at the Lompoc Valley Historical Society.

26. "Overflow Crowd Attends USO Club Dedication," *Lompoc Record*, August 7, 1942, 1, 10; "Ac-

tivities for USO Penthouse Are Outlined," *Lompoc Record*, July 31, 1942, 8; USO booklet "Dedication Program," August 2, 1942, Lompoc Valley Historical Society.

27. "Glamour Girls Out as USO Club Hostesses," *Lompoc Record*, October 2, 1942, 5.

28. "All-Star Show at Arena Next Week," *Camp Cooke Clarion*, July 17, 1942, 1; United Service Organizations, "Camp Shows Publicity Records, 1941–1957," microfilm reel No. 1, 958, Margaret Herrick Library, Academy of Motion Picture Arts and Sciences, Beverly Hills, CA.

29. "'Crazy Show' Comes Here Monday," *Camp Cooke Clarion*, August 21, 1942, p. 1; "Crazy Show Program," *Camp Cooke Clarion*, August 21, 1942, 1, 4.

30. "Heifetz Thrills Service Club Audience at Brilliant Concert," *Camp Cooke Clarion*, October 16, 1942, 1; United Service Organizations, "Camp Shows Publicity Records, 1941–1957," microfilm reel No. 1, 965, Margaret Herrick Library, Academy of Motion Picture Arts and Sciences, Beverly Hills, CA.

31. "Esty, Morris Create Third Camel Troupe," *The Billboard*, March 27, 1943, 3; photo and caption of young girl distributing cigarettes, *Cooke Clarion*, April 5, 1946, 3.

32. "Big Camel Stage Show, Arena Next Tuesday," *Camp Cooke Clarion*, October 23, 1942, 1, 3; "Camel Violinist Plays with Teeth," *Camp Cooke Clarion*, October 23, 1942, 1.

33. "Kay Kyser Here Wednesday," *Camp Cooke Clarion*, November 27, 1942, 1–2; *Camp Cooke Clarion*, photo caption of Kay Kyser, December 11, 1942, 3.

34. "'Hollywood on Parade' Scores Second Big Show Month Success," *Camp Cooke Clarion*, December 11, 1942, 1.

35. "'Arsenic and Old Lace,' Laugh Hit, Plays Two Night Stand at Camp," *Camp Cooke Clarion*, December 25, 1942, 1; Otis Sheridan, Member of the Original Cast of 'Arsenic and Old Lace,' Plays Here," *Camp Cooke Clarion*, December 25, 1942, 3.

36. "'Flying Colors' Is Entertainment Hit," *Camp Cooke Clarion*, January 8, 1943, 1–2.

37. "Ada Leonard All-Girl Review Receives Enthusiastic Applauses," *Camp Cooke Clarion*, January 22, 1943, 1; "No Glamor [*sic*] in Soldier Shows," *Yank: The Army Weekly*, September 10, 1943, 21.

38. "Ada Leonarda [*sic*] Show on Mon.–Tues.," *Camp Cooke Clarion*, January 15, 1943, 1; "Ada Leonard All-Girl Review Receives Enthusiastic Applauses," *Camp Cooke Clarion*, January 22, 1943, 1.

39. "Hancock Group Thrills Cooke Music Lovers," *Camp Cooke Clarion*, January 29, 1943, 1.

40. "Champs Appear in Sports Show," *Camp Cooke Clarion*, February 5, 1943, 1, 6.

41. "Judy Is Laugh Riot as 'Junior Miss' Proves

Tops in Comedy," *Camp Cooke Clarion*, February 5, 1943, 1.

42. *Camp Cooke Clarion*, General Electric advertisement, February 12, 1943, 3; "'House of Magic' Show Parades Science Marvels," *Camp Cooke Clarion*, February 19, 1943, 1.

43. *Camp Cooke Clarion*, photo caption of Sgt. Joe Louis, March 5, 1943, 1; "'This Is the Army' Movie Scenes Filmed at Camp; Joe Louis Here," *Camp Cooke Clarion*, March 5, 1943, 5; Richard Bak, *Joe Louis: The Great Black Hope* (Dallas: Taylor Publishing, 1996), 222.

44. "Musical Hit, 'In the Groove,' Sat.–Sun.," *Camp Cooke Clarion*, February 26, 1943, 1, 3; "Fast Musical Revue, 'In the Groove,' Gets Soldier Praise," *Camp Cooke Clarion*, March 5, 1943, 1, 2, 6.

45. "New Theater No. 3 to Open Tonight," *Camp Cooke Clarion*, February 26, 1943, 1; "Bob Hope 'Pours It On' for 6,000 Cooke Soldiers in Coast-to-Coast Broadcast and Sports Arena Hit," *Camp Cooke Clarion*, March 5, 1943, 1.

46. "New Theater No. 3 to Open Tonight," *Camp Cooke Clarion*, February 26, 1943, 1; "Where to Go," *Camp Cooke Clarion*, March 5, 1943, 2.

47. "Col. Carle H. Belt Officially Opens Service Club No. 2; Lounge Room and Fountain Have New Features," *Camp Cooke Clarion*, March 12, 1943, 1–2.

48. "Refreshing 'Claudia' Plays to Capacity Crowds," *Camp Cooke Clarion*, March 19, 1943, 1.

49. "Camp Cooke Will Be on the Air Sunday," *Camp Cooke Clarion*, March 19, 1943, 1; "Second Radio Show on the Air Sunday," *The Clarion*, March 26, 1943, 1; Purkiss, 19, 49.

50. "Don Cossack Chorus Receives High Praise for Concerts Here," *The Clarion*, April 9, 1943, 1, 3.

51. "Division Men in 'Baptism of Fire' Film," *Cooke Clarion*, October 8, 1943, 3.

52. Ibid.; Internet Archive, *Baptism of Fire*, http://archive.org/details/Baptism of Fire (accessed December 19, 2012).

53. "Concert Again Scores Success," *The Clarion*, April 9, 1943, 4; *The Clarion*, photo caption, "Vocalist Entertains Camp Cooke Soldiers," April 23, 1943, 3.

54. "'Hellzapoppin' Is Screamlined Comedy Hit," *The Clarion*, April 23, 1943, 1.

55. "Gray Gordon Band Spotlights 'Jump Rhythm,'" *Cooke Clarion*, May 28, 1943, 3.

56. "Smash Roxy Revue Stars Colorful Chorus Cuties," *Cooke Clarion*, June 11, 1943, 3.

57. "5000 See Santa Ana Flyers Ground Post-Division Clubs in Tight Series," *Cooke Clarion*, June 18, 1943, 5.

58. "Camp to Celebrate Father's Day," *Cooke Clarion*, June 18, 1943, 1, 3; "Civilian Dads Enjoy Soldier Program Sunday," *Cooke Clarion*, June 25, 1943, 1, 4.

59. "Crosby Arranged Talent Show on Sports Arena Stage Saturday Night," *Cooke Clarion*, July

9, 1943, 1, 3; "Hollywood Brings Spicy Program," *Cooke Clarion*, July 16, 1943, 3.

60. "Camp-Shows Musical in Arena 16–17," *Cooke Clarion*, July 9, 1943, 1–2; "'Swing's the Thing' Scores in Arena Camp-Shows Revue," *Cooke Clarion*, July 23, 1943, 1.

61. "11 Depot Show 'City of Stars' Is Cooke Hit," *Cooke Clarion*, July 30, 1943, 1, 3.

62. "Metropolitan Star Mary Van Kirk Presents Recital," *Cooke Clarion*, August 6, 1943, 1.

63. "Recreation Center at Surf Railroad Station Formally Opened Sunday," *Cooke Clarion*, August 6, 1943, 1; "Surf USO Club Separate Soon Due to Growth," *Lompoc Record*, February 25, 1944, 1; "Surf USO Club Slate Official Opening Sunday," *Lompoc Record*, April 28, 1944, 1; Myra Manfrina, "Surf USO Now a Memory as 3-Year Service Ends," *Santa Barbara News-Press*, May 6, 1946, A-4.

64. Manfrina, A-4.

65. Ibid.

66. Ibid.

67. Ibid.; "Improvements at Surf USO to Start Soon," *Cooke Clarion*, July 6, 1945, 4.

68. "Hancock Ensemble Concert Brings Famed Mezzo-Soprano Thursday," *Cooke Clarion*, July 30, 1943, 1.

69. "Lockheed Lovelies Grace 559th Week-End Party in Camp," *Cooke Clarion*, August 27, 1943, 1–2.

70. "Camel Caravan Showing at Cooke Aug. 25 and 26," *Cooke Clarion*, August 20, 1943, 1–2; "Camel Caravan in Two Night Stand Makes Hit," *Cooke Clarion*, August 27, 1943, 3.

71. "Two Night Show to Play Here," *Cooke Clarion*, August 27, 1943, 1–2; "'Passing Parade' Brings Second Musical in Week," *Cooke Clarion*, September 3, 1943, 3.

72. "Jane Wyman Guest of Hq. Btry., 280 FA," *Cooke Clarion*, September 3, 1943, 1.

73. "Warner Brothers Will Film Filipinos Sunday," *Cooke Clarion*, September 3, 1943, 1; "30 Filipino Soldiers and 6 Young Women Play Roles in Film," *Cooke Clarion*, September 10, 1943, 1, 4; obituary for Pacita Todtod Bobadilla, Dignity Memorial—Lima Family Santa Clara Mortuary, http://www.dignitymemorial.com/en-us/obituary (accessed March 11, 2013).

74. "El Paseo Revue at Sports Arena," *Cooke Clarion*, September 3, 1943, 1; "Grand International Revue Pleases GIs," *Cooke Clarion*, September 10, 1943, 3.

75. "Los Angeles USO Lovelies Visit Sixth," *Cooke Clarion*, September 17, 1943, 4.

76. Ibid.

77. "Camp Show, WLS National Barn Dance, Monday and Tuesday," *Cooke Clarion*, October 22, 1943, 1, 3; "WLS Barn Dance Camp Show in Two Night Stand Proves Good Entertainment," *Cooke Clarion*, October 29, 1943, 3.

78. "Preston Foster Visit Here Hits New High," *Cooke Clarion*, October 29, 1943, 1, 3.

79. "30 Girls Will Be in Revue at Both Clubs on Sunday," *Cooke Clarion*, November 5, 1943, 1; "Faye Porter Show Another Hit at Clubs," *Cooke Clarion*, November 12, 1943, 2.

80. "Joe Louis Boxing Exhibition Set for Friday Night, November 12," *Cooke Clarion*, November 5, 1943, 1; "Joe Louis Exhibition Only Fair' 'Sugar' Robinson Cops Spotlight," *Cooke Clarion*, November 19, 1943, 4–5.

81. "Joe Louis Boxing Exhibition Set for Friday Night, November 12," *Cooke Clarion*, November 5, 1943, 1; "Joe Louis Exhibition Only Fair' 'Sugar' Robinson Cops Spotlight," *Cooke Clarion*, November 19, 1943, 4–5.

82. "'Good, Not So Good,' Labels 'Showing Off,'" *Cooke Clarion*, November 12, 1943, 3.

83. "Durante Touch Puts Over Show," *Cooke Clarion*, November 19, 1943, 1, 3; "Jimmy Durante Tops at Club 2 Sunday Evening," *Cooke Clarion*, November 19, 1943, 3.

84. "Durante Touch Puts Over Show," *Cooke Clarion*, November 19, 1943, 1, 3.

85. "Ten Thousand Khaki Clad Lads See Kay Kyser Shows in Arena," *Cooke Clarion*, November 19, 1943, 1, 5.

86. "'Let's Go' Is Sports Arena Show Next Monday-Tuesday at 1930," *Cooke Clarion*, November 19, 1943, 1, 3; "Colored Revue USO-Camp Show Sends," *Cooke Clarion*, November 26, 1943, 3.

87. "'Going Some' Is 'Fairly' Fast Revue," *Cooke Clarion*, December 10, 1943, 3.

88. "20th Century–Fox Using Division in Filming 'I Married a Soldier,'" *Cooke Clarion*, December 10, 1943, 1, 6; American Film Institute Catalog, *In the Meantime, Darling*, http://afi.chadwick.com/film/full_rec?action=BYID&FILE=///session/ (accessed April 30, 2010).

89. "'Furlough Fun' Proves Hilarious Show," *Cooke Clarion*, December 24, 1943, 1; Maxene Andrews and Bill Gilbert, *Over Here, Over There: The Andrews Sisters and the USO Stars in World War II* (New York: Zenith Books, 1993), 113.

90. "New USO Hit Show in Arena," *Cooke Clarion*, December 17, 1943, 1, 4; "'Come What May,' Latest Camp Show, Has GIs Open Mouth with Buddies on Stage in Trance," *Cooke Clarion*, December 24, 1943, 4.

91. "Weekend Holiday Show Rated Super," *Cooke Clarion*, December 31, 1943, 3.

92. John Strategakis, "31st Medics Entertain Desert Battalion," *Cooke Clarion*, January 7, 1944, 3.

93. "'Desert Battalion' to Aid Celebration of 28th AAA Birthday," *Cooke Clarion*, March 10, 1944, 3; "28th AAA Group Host to Members of Desert Battalion on Weekend," *Cooke Clarion*, March 17,1944, 8.

94. Jack Preston and Mrs. Edward G. Robinson, *The Desert Battalion* (Hollywood, CA: Murray & Gee, 1944), 1, 6.

95. Ibid., 6, 26, 59.

96. "Curvesome Paula Peiry Steals Camp Shows Hit 'What's Next,'" *Cooke Clarion*, January 19, 1944, 4.

97. "'Shell Show,' Variety Hit, to Play Here Two Nights," *Cooke Clarion*, January 21, 1944, 1, 3; "GIs 'Hash-Over' Shell Show, Like Lucille Elmore Best," *Cooke Clarion*, January 28, 1944, 3.

98. "Lt. Rudy Vallee and Coast Guard Band in Concert as Bond Sale Up," *Cooke Clarion*, February 4, 1944, 1, 3.

99. "'Say When' Plays Two Night Stand," *Cooke Clarion*, February 11, 1944, 3.

100. "Star-Lined Hollywood Revue in Arena Sunday," *Cooke Clarion*, February 11, 1944, 1, 3; "Collinstone Revue Sunday Proves Top Entertainment," *Cooke Clarion*, February 18, 1944, 1, 3.

101. "Wonders of Electricity Displayed at 'House of Magic' Arena Show," *Cooke Clarion*, February 25, 1944, 3.

102. "Ferde Grofe to Head Service Club Shows by Hancock College," *Cooke Clarion*, February 18, 1944, 1; "Pianist-Composer Ferde Grofe, Featured at Weekend Club Shows," *Cooke Clarion*, March 3, 1944, 3.

103. "Revue to Play in Arena," *Cooke Clarion*, March 3, 1944, 1, 3; "Brucettes, Impressionist Layne Tope Above Average USO Show," *Cooke Clarion*, March 10, 1944, 3.

104. "'Stars of Tomorrow,' Hollywood Stage Show, to Be Here Sunday," *Cooke Clarion*, March 24, 1944, 1, 6; "'Stars of Tomorrow,' Mom Fletcher Show, Produce Plenty GI Whistles," *Cooke Clarion*, March 31, 1944, 3.

105. "USO Camp Show, 'Funny Side Up,' to Play Sports Arena Stage Wednesday-Thursday," *Cooke Clarion*, March 31, 1944, 1; "'Funny Side Up' Is Solid Sender," *Cooke Clarion*, April 7, 1944, 3.

106. "Lt. Valle, Band, Here Monday," *Cooke Clarion*, April 21, 1944, 1; "Lt. Rudy Vallee Versatile Band, Tremendous Hit," *Cooke Clarion*, April 28, 1944, 1, 3.

107. "'Around the Corner,' USO Revue on Sports Arena Stage May 4–5," *Cooke Clarion*, April 28, 1944, 1, 4.

108. "Famed Glove Champions Referee Bouts for Thunderbolt Ring Team," *Cooke Clarion*, June 9, 1944, 5; "Bug Eyed Youths See Arena Greats at Camp Cooke," *Lompoc Record*, June 9, 1944, 1.

109. "New USO Hit Show in Arena June 8–9," *Cooke Clarion*, June 2, 1944, 1, 4; "'All Is Well' Proves Good Entertainment in Two-Night Arena Run," *Cooke Clarion*, June 9, 1944, 1–2.

110. "'Clambake Follies' Coming to Arena Stage Sunday," *Cooke Clarion*, June 16, 1944, 1, 3; "'Clambake Follies' Plays to Packed Arena," *Cooke Clarion*, June 23, 1944, 1, 3.

111. "Fortunio Bonanova, Hollywood Singing Star, Will Make Personal Appearance at Cooke," *Cooke Clarion*, June 23, 1944, 1, 3; "Latin Singing

Star Entertains Patients," *Cooke Clarion*, June 30, 1944, 3.

112. "Next USO Show Due July 10–11," *Cooke Clarion*, July 7, 1944, 1, 3; Pfc. Charles A. Mitchell, "'Brazilian Nights' Scores with GIs in Two Shows," *Cooke Clarion*, July 14, 1944, 3.

113. "Orson Wells Brings Show," *Cooke Clarion*, July 14, 1944, 1, 3.

114. "'What's Cookin', Coming Here Aug. 9–10 Solid Colored Revue," *Cooke Clarion*, August 4, 1944, 1; Charles A. Mitchell, "Harlem Revue Entertains GIs," *Cooke Clarion*, August 11, 1944, 1, 4.

115. "Hollywood Show Packs in 7,000," *Cooke Clarion*, August 18, 1944, 1, 3.

116. "Jimmy Dorsey Band Makes Hit," *Cooke Clarion*, September 8, 1944, 1, 3.

117. Charles A. Mitchell, "Charlie Withers Proves Big Laugh Provoker in Latest USO Show," *Cooke Clarion*, September 15, 1944, 3.

118. "Lively Variety Revue Scores Big Hit with GIs," *Cooke Clarion*, October 13, 1944, 1, 3.

119. "Johnny O'Brien Takes Top Show Spot," *Cooke Clarion*, November 3, 1944, 3.

120. S. D. Johnnie, "Better Than Usual USO Show, 'Thanks Loads,' Produces Laughs," *Cooke Clarion*, November 10, 1944, 3; "'Thanks Loads' to Give Comedy, Pretty Girls, Hit Tunes at Arena," *Cooke Clarion*, November 3, 1944, 1, 6.

121. "Tankers, Depotmen in Film of Stalingrad Fight," *Cooke Clarion*, December 1, 1944, 1, 3.

122. Ibid.

123. Ibid.; "Red Army Tank Battle Film Retakes Made," *Cooke Clarion*, December 8, 1944, 1.

124. Charles A. Mitchell, "GI Talent Fills in for USO Show," *Cooke Clarion*, December 8, 1944, 3.

125. Ibid.

126. "'Star Spangled Circus' to Open One Week Run in Arena Sunday," *Cooke Clarion*, December 8, 1944, 1, 6; "Star Spangled Circus in Finale Saturday Night," *Cooke Clarion*, December 15, 1944, 1, 3; Neil Cockerline, Circus History Database, to author, "3106 Benny Fox Circus, La Tosca," http://www.circushistory.org/Query.htm (accessed October 15, 2009).

127. "Our Hat's Off to 'Hats Off' Show," *Cooke Clarion*, January 19, 1945, 1, 3.

128. "MC Jack Leonard Hit of USO Show Monday," *Cooke Clarion*, February 16, 1945, 1, 4.

129. F. Scott Wilder, "'Perk Up,' USO Show Pleases GI Audience," *Cooke Clarion*, April 13, 1945, 1, 4.

130. "'Broadway Maneuvers' at Arena Thursday," *Cooke Clarion*, May 4, 1945, 1; Scott Wilder, "OK Label Put on USO Show," *Cooke Clarion*, May 11, 1945, 1, 4.

131. "Hello Joe' USO-Camp Show, Will Play Sports Arena, Friday June 9," *Cooke Clarion*, June 1, 1944, 1; "'Hello Joe,' USO Show, Wins Audience Okay," *Cooke Clarion*, June 15, 1945, 3.

132. "'Chicks and Chuckles,' July USO Show, Plays Arena Tuesday Night," *Cooke Clarion*, July 6, 1945, 1; George Spellman, "Few 'Chicks' 'n No Chuckles,'" *Cooke Clarion*, July 13, 1945, 2.

133. "'Gee-Eye Revue' Plays Arena Sat.," *Cooke Clarion*, August 10, 1945, 1; "Reviewer Labels USO Show 'NG,'" *Cooke Clarion*, August 17, 1945, 3.

134. "USO Show, 'Have a Look,' Slated for Two Performances in Sports Arena Monday Eve," *Cooke Clarion*, September 7, 1945, 1, 3; "Latest USO Show Way Above Average," *Cooke Clarion*, September 14, 1945, 4.

135. "Larry Crosby to Bring 'Clambake Follies' Here," *Cooke Clarion*, September 7, 1945, 1, 3; "5000 GIs Roar Approval for Clambake Follies," *Cooke Clarion*, September 14, 1945, 3.

136. "Hollywood 'Spot Show' Wins Over 8,000 Cheers," *Cooke Clarion*, September 21, 1945, 3.

137. "Louis Armstrong in Arena Monday," *Cooke Clarion*, September 21, 1945, 1, 3; "'Satchmo's' Sock Sends 'em," *Cooke Clarion*, September 28, 1945, 3; "Negro Band 'Loaded' with Colorful Figures," *Cooke Clarion*, September 28, 1945, 3.

138. "Latest Hollywood 'Spot Show' Has Variety," *Cooke Clarion*, October 5, 1945, 3.

139. "Two Bob Hope Shows 'Hilarious Hits,'" *Cooke Clarion*, October 12, 1945, 1, 3.

140. "Park Avenue 'Debs' USO Show Hit," *Cooke Clarion*, October 12, 1945, 4.

141. "Spot Show Spiked with Spunk 'n Spirit," *Cooke Clarion*, October 19, 1945, 1, 3; "Entertainer Dies During Camp Show," *Los Angeles Times*, October 16, 1945, A7.

142. Willys Peck, "Shell Show Success Added to Recent Cooke Entertainment Hits," *Cooke Clarion*, October 26, 1945, 3.

143. Willys Peck, "Carson Comic 'Gives' with Hit Variety Show," *Cooke Clarion*, November 9, 1945, 1, 3.

144. "Sid's Show Survives Stiff Test," *Cooke Clarion*, November 16, 1945, 3.

145. "Short, Snappy USO Show, 'You Said It,' Gets 'OK' Not from Critical Audience," *Cooke Clarion*, November 30, 1945, 6.

146. "Wilde Twins, Chill Wills Head Latest Spot Show," *Cooke Clarion*, November 30, 1945, 3.

147. "1,500 Turn Out for USO Show," *Cooke Clarion*, December 28, 1945, 1, 4.

148. "Kiss and Tell,' Hilarious Broadway Stage Play, Sponsored by USO-Camp Show, Coming to Sports Arena January 31," *Cooke Clarion*, January 25, 1946, 1, 3; "Enthusiastic Audience Greets 'Kiss and Tell,'" *Cooke Clarion*, February 1, 1946, 1.

149. "Laughs, Whistles Greet USO Spot Show on Tuesday Night," *Cooke Clarion*, February 15, 1946, 1, 4; http://www.hillbilly-music.com (accessed January 25, 2013).

150. "USO Show Due Here March 8," *Cooke Clarion*, March 1, 1946, 1; "USO Show with 12-Piece Orchestra Is Hailed by 500 GIs," *Cooke Clarion*, March 15, 1946, 1, 3.

151. "Shell Military Show Is Given Huge

Hand; Johnny O'Brien Proves Audience Favorite," *Cooke Clarion*, April 5, 1946, 3.

152. "H Street USO Club Closing Now Definite," *Lompoc Record*, June 1, 1945, 1; "Landmark of City Set for Demolition," *Lompoc Record*, December 10, 1959, 1.

153. "Santa Maria USO Closes March 31," *Santa Maria Daily Times*, March 15, 1946, 1–2; "Surf USO Club Closes After Three Years of Wartime Service," *Lompoc Record*, May 2, 1946, 1, 6; "USO Buildings at Surf Being Sold by Army," *Lompoc Record*, September 23, 1948, 1; "USO Program for Community This Week-End," *Lompoc Record*, May 23, 1946, 1, 12 (latter page unnumbered); "Final USO Vote Carries by 3 to 2," *Lompoc Record*, August 7, 1947, 1, 8; "Center Officially Opened in Formal Dedication," *Lompoc Record*, September 11, 1947, 1, 4; "Formal Delivery of USO Building Is Completed," *Lompoc Record*, October 23, 1947, 1.

154. "'Welcome' Mat Put Out at City Center, Soldiers Get Priority," *Lompoc Record*, September 14, 1950, I-7.

155. "'Welcome' Mat Put Out at City Center, Soldiers Get Priority," *Lompoc Record*, September 14, 1950, I-7; "Square Dancers to Teach Camp Cooke Soldiers," *Lompoc Record*, September 28, 1950, II-2; "223rd Regiment Band to Play at Center Dance," *Lompoc Record*, January 18, 1951, 2; "AWVS Plans Christmas Entertainment for Armed Forces at Center and Camp," *Lompoc Record*, December 20, 1951, 1.

156. "City 'Donates' Recreation Bldg. for USO's Use," *Santa Maria Times*, February 20, 1951, 1, 8; "Old City Recreation Center Made Like New for USO Center," *Santa Maria Times*, February 21, 1951, 1; "USO to Serve as Peace Time Agency," *Santa Maria Times*, March 1, 1951, 1, 6; "GI Guide...," *Santa Maria Times*, February 22, 1952, 1; "YMCA Lands Building Site," *Santa Maria Times*, January 16, 1956, 1–2.

157. "Welcome 40th Lompoc Says," *Lompoc Record*, September 7, 1950, 1.

158. "Ralph Edwards' Radio Show Booked at Cooke," *Lompoc Record*, September 21, 1950, 5; "Official Diary," 9–10.

159. "All-Star Show Set Tomorrow at Camp Cooke," *Santa Maria Times*, November 7, 1950, 8.

160. "Official Diary," 25. Marjorie Hall was a well-known dance instructor in Santa Maria. On March 17, 1966, she was struck by an automobile while crossing at the intersection of Broadway and Morrison on her way to an opera at Santa Maria High School. The 59-year-old Hall slipped into a coma and died ten days later. In 1967, a magnificent fountain and pond was built at the corner of Broadway and Main and dedicated to her memory. The memorial plaque says, "She had devoted her life to the children of Santa Maria, teaching them to dance and to appreciate the fine arts."

161. Ibid., 32.

162. *Cooke Clarion*, photo caption of Phil Regan, March 21, 1951, 5; Hollywood Coordinating Committee Talent Records, Spot Show No. 2628, March 1951, 5, Margaret Herrick Library, Academy of Motion Picture Arts and Sciences, Beverly Hills, CA.

163. The date and place for this show is listed on the photos at the California State Military Museum.

164. "Tonight's Show Should Be the Best Yet," *Cooke Clarion*, April 27, 1951, 2; "Star-Studded Shows Scheduled for Cooke Hospital Audiences," *Cooke Clarion*, April 27, 1951, 4.

165. Hollywood Coordinating Committee Talent Records, May 1951, 2, Margaret Herrick Library, Academy of Motion Picture Arts and Sciences, Beverly Hills, CA.

166. Internet Movie Database, "Biography for Elinor Donahue," http://www.imbd.com/name/nm0231942/bio (accessed December 12, 2012).

167. Letter from Elinor Donahue to the author, December 4, 2012.

168. Ibid.

169. Ibid.

170. Ibid.

171. Ibid.

172. Ibid.

173. Roy Bowman, "Beck, Brown and White Cut Colorful Capers," *Cooke Clarion*, June 8, 1951, 4.

174. "Bergen Entertains Camp Cooke Soldiers," *Santa Maria Times*, June 23, 1951, 1.

175. Ray Bowman, "Pasadena Players Score Big Hit with Brilliant Comedy, Roadside," *Cooke Clarion*, July 27, 1951, 4; Ray Bowman, "Circle Theater Troupe Shows There's Spice in Life," *Cooke Clarion*, August 10, 1951, 4.

176. "Marvin Miller and Starlet Keep Sports Arena Audience Enthralled," *Cooke Clarion*, August 17, 1951, 4.

177. "Bob Hope to Entertain at Cooke Tonight," *Lompoc Record*, April 10, 1952, II-8; "Official Diary," 104.

178. "Official Diary," 132.

179. Ibid., 148.

180. Ibid., n.p.

181. "Gala Carnival at Camp Cooke Open," *Lompoc Record*, October 30, 1952, 1, 8; "Rides Are Obtained for Cooke Fete," *Lompoc Record*, October 16, 1952, 1; program booklet *Camp Cooke Welfare Carnival*, October 1952, n.p.

182. "Hit Scored Here by Harry Owens' Royal Hawaiians," *Cooke Clarion*, December 3, 1952, 4.

183. "History of the Santa Maria Public Library," on file in the library archives, updated June 16, 2009.

## Chapter 6

1. "Expect No Interference in Training," *Lompoc Record*, November 13, 1952, 1, 8; Brig.

Gen. Stuart H. Sherman, Jr., to CINCSAC/DE, September 8, 1975, History Office, 30th Space Wing, Vandenberg AFB, CA.

2. "Union Makes Oil Strike on Cooke," *Lompoc Record*, September 25, 1952, 1; "Cooke Decision Stirs Up Furor," *Santa Maria Times*, November 8, 1952, 1, 8; Lee Keeling and Associates, *First Gross Mineral Appraisal* [of] *Vandenberg Air Force Base, Santa Barbara County, California* (Tulsa, OK: 1977), 5, schedules 2 and 3, n.p.

3. "Camp Cooke to Close in Month," *Santa Maria Times*, November 7, 1952, 1; "Cooke Decision Stirs Up Furor," *Santa Maria Times*, November 8, 1952, 1, 8; "Gov. Warren Raps Cooke Closure Move," *Santa Maria Times*, November 15, 1952, 1; Bill Phillips, "Kirk to Take Area Case Before Powers in D of C," *Santa Maria Times*, November 11, 1952, 1, 8; Carroll Gewin, "Drive to Keep Cooke Rolling," *Santa Maria Times*, November 13, 1952, 1, 8; "1,000 Jobs Set for Cooke Force," *Santa Maria Times*, November 20, 1952, 1; "Camp Cooke Inactivation Controversy," editorial, *Lompoc Record*, November 27, 1952, II-8.

4. "$4.3 Million Annually to Be Saved at Cooke—D of A," *Santa Maria Times*, November 19, 1952, 1; "Three Other Army Camps to Be Closed," *Santa Maria Times*, November 15, 1952, 1; "Army Tells Reasons for Closing Camp," *Lompoc Record*, November 20, 1952, I-1, 8; "Army Revels Cooke Repairs Cost $8.5 Million," *Lompoc Record*, November 29, 1952, I-1.

5. "Oil Activity Isn't Bar to Army—Union," *Santa Maria Times*, November 13, 1952, 1, 8.

6. Lee Keeling and Associates, 5, schedules 2 and 3, n.p; Sherman, Jr., to CINCSAC/DE, September 8, 1975; Col. Charles F. Wilhelm, WSMC/CC to Maj. Gen James H. Marshall, SAMTO/CC, "AF Decision to Allow Mineral Rights Subordination Agreements to Lapse," June 23, 1980, History Office, 30th Space Wing, Vandenberg AFB, CA.

7. "Camp Cooke Inactivation Controversy," *Lompoc Record*, November 27, 1952, II-8.

8. "First 44th Unit Bound for Lewis in Motor Convoy," *Santa Maria Times*, November 26, 1952, 1; "1,500 44th to Fort Lewis," *Santa Maria Times*, December 11, 1952, 1.

9. "Cooke's Hospital Officially Closed," *Santa Maria Times*, March 2, 1953, 1; "Cooke Clarion Swan Song Due," *Santa Maria Times*, December 16, 1952, 1; "Public Auction, Camp Cooke—Dec. 20," *Santa Maria Times*, advertisement, December 18, 1952, 8.

10. "Last Units Quit Cooke Monday A.M.," *Santa Maria Times*, January 2, 1953, 1; "Inactivation of Cooke Proceeding," *Lompoc Record*, December 18, 1952, I-1, 3.

11. "Ceremony 'Inactivates' Camp Cooke," *Santa Barbara News-Press*, April 1, 1953, A-4; "Ceremonial Flag Lowering Marks Camp Cooke Closing," *Santa Maria Times*, April 1, 1953, 1. Page

two of the article is missing from the microfilm copy of the newspaper at the Santa Maria library.

12. "Cooke Flag to Go Down Last Time," *Lompoc Record*, March 26, 1952, 1, 8.

13. "Cattle Move Onto Camp Cooke," *Lompoc Record*, April 2, 1953, 1; "Cooke Lease to Bring Near $1 Million," *Lompoc Record*, October 29, 1953, 1; Eve Allgair, "Bleats Replace the Bugles at Camp Cooke Site as 2,000 Unregimental Sheep Take Over Area," *Santa Barbara News-Press*, April 2, 1953, B-2.

14. "DB Releases Puerto Ricans," *Lompoc Record*, July 16, 1953, 1; "Puerto Rican 'Cowards' to Cooke," *Santa Maria Times*, March 3, 1953, 1; Gilberto N. Villahermosa, *Honor and Fidelity: The 65th Infantry in Korea, 1950–1953* (Washington, D.C.: Center of Military History, U.S. Army, 2009), 268–70.

15. "L.A. Firm Is Low Barracks Bidder," *Lompoc Record*, October 21, 1954, 1; "General Dean Tours USDB," *Lompoc Record*, January 27, 1955, 1; "Moving Day Comes to Branch DB," *Lompoc Record*, May 9, 1956, 1; "Lt. Col. Klein Succumbs Here Friday," *Lompoc Record*, September 29, 1955, 1–2.

16. Construction Starts on DB Housing," *Lompoc Record*, May 10, 1956, 1; "Army Families Move on Post," *Lompoc Record*, February 21, 1957, 6; "DB to Get 60 Capehart Home Units," *Lompoc Record*, March 13, 1958, 1; "Status of DB Housing Clarified," *Lompoc Record*, June 19, 1958, 8d; "Recreation Area Open at Cooke," *Lompoc Record*, July 5, 1956, 1; "USDB Chapel Dedicated on Sunday," *Lompoc Record*, July 25, 1957, 1–2; "Construction Started on USDB Gym," *Lompoc Record*, February 28, 1957, II-1.

17. "Prison to Operate at Capacity," *Lompoc Record*, May 28, 1959, 1; "USDB Processed 20,608 Prisoners in 12 Years," *Lompoc Record*, May 28, 1959, 1–2; "Confinement Facilities of DB Shift to Gov't," *Lompoc Record*, August 3,1959, 1. It is now a Federal Correctional Institution.

## *Chapter 7*

1. Frederick I Ordway III and Mitchell R. Sharpe, *The Rocket Team* (New York: Thomas R. Cromwell, 1979), 186, 195, 196.

2. Jacob Neufeld, *The Development of Ballistic Missiles in the United States Air Force, 1945–1960* (Washington, D.C.: Office of Air Force History, U.S. Air Force, 1990), 43–44, 48.

3. Neufeld, 94, 97, 102; Mark C. Cleary, *The 6555th: Missile and Space Launches Through 1970*, 45th Space Wing History Office, Patrick AFB, FL, November 1991, 82.

4. Neufeld, 121, 144, 146, 241.

5. Ibid., 201; Cleary, 82, 83.

6. MSgt. Max Gerster, *History of the 1st Missile Division and the 392nd Air Base Group, 15 April 1957 Through 30 June 1957*, Cooke AFB, CA, 1957, attached document No. 2, 58–64, SWHO.

7. Gerster, 65, 72.
8. Ibid., 16; "Permit to the Department of the Air Force," signed by Col. Arthur H. Frye, Jr., Army Corps of Engineers, January 25, 1957, SWHO; "Air Force Vanguard Now Stationed at Camp Cooke," *Lompoc Record,* January 24, 1957, 1–2. Smith arrived at Cooke on December 1, 1956. Shaneyfelt probably arrived at or about the same time. The Army's disciplinary barracks provided the new Air Force arrivals with meals and housing until these services became available on the air base.
9. Gerster, 17, 18, chronology, n.p.; MSgt. Max Gerster, *History of the 1st Missile Division, 1 July Through 31 December 1957,* Cooke AFB, CA, 1958, xi, SWHO.
10. Historical Division, *History of Air Research & Development Command, 1 July–31 December 1960,* Vol. III, 1961, 4–5, SWHO. This report incorrectly lists the groundbreaking as occurring on May 8.
11. Ibid., 4; Gerster, *History of the 1st Missile Division, 1 July Through 31 December 1957,* 134–35; Gerster, *History of the 1st Missile Division and the 392nd Air Base Group, 15 April 1957 Through 30 June 1957.* See chronology, n.p.
12. *History of Air Research & Development Command, 1 July–31 December 1960,* 6–7; Carl Berger, *History of the 1st Missile Division,* Vandenberg AFB, CA, n.d., 22–23, SWHO. The homes were named for Republican Senator Homer Capehart, who in 1955 sponsored a bill for new military family housing that was enacted into law. Between 1996 and 2012 the Air Force replaced the first-generation housing at Vandenberg with new homes, and presently has just under 1,000 units.
13. Berger, *History of the 1st Missile Division,* 23.
14. Ibid., 27; Robert J. Watson, *Into the Missile Age, 1956–1960,* Vol. IV of *History of the Office of the Secretary of Defense,* ed. Alfred Goldberg (Washington, D.C.: Historical Office, Office of the Secretary of Defense, 1997), 388.
15. Space Division History Office, *Space and Missile Systems Organization: A Chronology, 1954–1979,* Los Angeles AFB, CA, n.d., 46–48; Berger, *History of the 1st Missile Division,* 29–30, 33.
16. *Space and Missile Systems Organization: A Chronology, 1954–1979,* historical report; "6565th Test Wing (Development), 20 October 1960 Through 31 December 1960," Vandenberg AFB, n.d., SWHO. The Field Office evolved into the 6595th Test Wing in October 1960. In 1961, ARDC was renamed Air Force Systems Command.
17. Historical narrative and schedule of events for dedication of Vandenberg Air Force Base, n.d., SWHO.
18. Arthur Menken, *History of the Pacific Missile 1 November 1945 to 30 June 1959,* U.S. Navy, Point Mugu, CA, n.d., 4–5 and appendix A, 4–5; "Pacific Missile Range," presentation by Rear Admiral Jack P. Monroe, commander of the Pacific

Missile Range, to Walker L. Cisler, October 1959, 7–8, 10–11, SWHO; anonymous background paper "Background of Vandenberg/Arguello Relationships," n.d., SWHO; "Preliminary Survey Is Made Here," *Lompoc Record,* July 26, 1956, 1; "Engineers Silent on Cooke Job," *Lompoc Record,* September 13, 1956, 1.
19. Menken, appendix A, 4–5, 7; J. W. Caldwell, *Command History 1960, A Historical Report: U.S. Naval Missile Facility Point Arguello,* U.S. Navy, Point Mugu, CA, 1962, 1, 2–3.
20. James Baar and William E. Howard, "AF-Navy Space Range Fight Nears Showdown," *Missiles and Rockets,* December 21, 1959, 11–14; Carl Berger, MSgt. Max Gerster, TSgt. Bruce Pollica and TSgt. Thomas N. Thompson, *History of the 1st Missile Division, 1 July–31 December 1959,* Vandenberg AFB, CA, 1960, 120–129, SWHO; Thomas Mendham "Whole Town Takes Off Every Time a Missile Takes Off," *National Enquirer,* October 27, 1963, 28; Jeffrey Geiger, "Vandenberg Launch Summary," n.d., SWHO.
21. James Baar and William E. Howard, "AF Attacks 'Secret' Navy Space Plan," *Missiles and Rockets,* January 18, 1960, 11; "Navy Denies Air Force's PMR Charges," *Missiles and Rockets,* January 25, 1960, 26; anonymous background paper "Background of Vandenberg/Arguello Relationships," n.d., SWHO; Watson, 189.
22. Robert D. Bickett, *History of the Air Force Western Test Range, 15 May–31 December 1964,* Vandenberg AFB, CA, n.d., 40–42, SWHO.
23. Caldwell, 8; Amendment No. 1 to Burke/LeMay agreement, December 31, 1960, SWHO; Watson, 386.
24. Memorandum for the Armed Services, Robert S. McNamara, "Management and Organization of DoD Ranges and Flight Test Facilities," November 16, 1963, SWHO; News release, "PMR-AFWTR Transfer of Range Operating Responsibilities," U.S. Air Force, Vandenberg AFB, February 2, 1965, SWHO. In 1979 the AFWTR name was shortened to Western Test Range.
25. Curtis Peebles, *High Frontier: The United States Air Force and the Military Space Program,* U.S. Air Force History and Museum Program, 1997, 22–23. Cape Kennedy was a temporary alias for Cape Canaveral.
26. Ibid., 23; Robert D. Bickett, *History of the Air Force Western Test Range, 1 January–31 December 1966,* Vandenberg AFB, n.d., 43–45, SWHO. The senate committee on Aeronautical and Space Sciences held hearings on February 24, 1966, to get an explanation of why a new Titan III-C launch facility was required at Vandenberg AFB for the Manned Orbiting Laboratory. Not more than a hint of the program's true objective can be extrapolated from these hearings.
27. Joseph T. Donahue, Jr., and Dean J. Stevens, *History of the 1st Strategic Aerospace Division, January–June 1966,* Vandenberg AFB, n.d., 225–26,

248    Notes—Chapter 7

SWHO; "$9 Million Payment for Sudden Ranch," *Santa Maria Times*, December 21, 1968, 10.

28. Bickett, *History of the Air Force Western Test Range, 1 January–31 December 1966*, 53–54; Robert D. Bickett, *History of the Air Force Western Test Range, 1 January–30 June 1969*, Vandenberg AFB, n.d., 54–55, SWHO; Peebles, 25–26; booklet "Reusable Launch Vehicle/Spacecraft Concept," Report No. GDC-DCB-68-017, General Dynamics Convair Division, November 1968.

29. "7-Day Report, WS-314A Demonstration Flight, Thor Missile 151," Douglas Aircraft Company, December 24, 1958, 1, SWHO; Geiger, "Vandenberg Launch Summary," 1, 7; Julian Hartt, *The Mighty Thor: Missile in Readiness* (New York: Duell, Sloan and Pearce, 1961), 213.

30. "7-Day Report, WS-314A Demonstration Flight, Thor Missile 151," December 24, 1958, 17; letter from John C. Bon Tempo to the author, April 29, 2006.

31. Geiger, "Vandenberg Launch Summary," 1, 7; Association of Air Force Missileers, with David K. Stumpf, *Air Force Missileers* (Paducah, KY: Turner Publishing, 1998), 23.

32. Kevin C. Ruffner, ed., *Corona: America's First Satellite Program* (Washington, D.C.: Central Intelligence Agency, 1995), xiii–xv, 12, 13, 16.

33. Ibid., 16–17.

34. Ibid., xiii–xv, 21–22, 24, 29, 38; William E. Burrows, *Deep Black: Space Espionage and National Security* (New York: Random House, 1986), 136–37.

35. Carl Berger, MSgt. Max Gerster, TSgt. Bruce Pollica, and TSgt. Thomas N. Thompson, *History of the 1st Missile Division, 1 January–30 June 1959*, Vandenberg AFB, CA, 1960, 77, SWHO.

36. Carl Berger, MSgt. Max Gerster, TSgt. Bruce Pollica, and TSgt. Thomas N. Thompson, *History of the 1st Missile Division, 1 July–31 December 1959*, Vandenberg AFB, Ca., 1960, 79–80, SWHO; Carl Berger and TSgt. Bruce Pollica, *History of the 1st Missile Division, 1 January–30 June 1960*, Vandenberg AFB, CA, 1960, 13n, SWHO.

37. Carl Berger and TSgt. Bruce Pollica, *History of the 1st Missile Division, 1 January–30 June 1960*, Vandenberg AFB, CA, 1960, 13, 13n, SWHO.

38. Carl Berger, Warren S. Howard, and SSgt. Ray A. Hanner, Jr., *History of the 1st Missile Division, 1 January–30 June 1961*, Vandenberg AFB, CA, 1961, 66–67, SWHO.

39. Geiger, "Vandenberg Launch Summary," n.p.

40. 6595th Aerospace Test Wing, summary report for Titan VS-1, "Silver Saddle," May 3, 1961, SWHO. Sometime after the 1960s the title missile flight safety officer was changed to missile flight control officer (MFCO).

41. Geiger, "Vandenberg Launch Summary," n.p.; 6595th Aerospace Test Wing, summary report for Titan SM 68-2, "Big Sam," September 23, 1961, SWHO.

42. NASA Historical Staff, *Astronautics and Aeronautics, 1963: Chronology on Science, Technology, and Policy* (Washington, D.C.: National Aeronautics and Space Administration, 1964), 57; 6595th Aerospace Test Wing, summary report for Titan II XSM68B-N8, "Dinner Party," April 27, 1963, SWHO.

43. Geiger, "Vandenberg Launch Summary," n.p.; Robert D. Bickett, *Space and Missile Test Center Index of Missiles Launched on the Western Test Range, 16 December 1958–31 December 1969*, Vandenberg AFB, CA, n.d., 30, SWHO.

44. U.S. House of Representatives, Committee on Science and Astronautics, *Astronautical and Aeronautical Events of 1962*, report prepared by the National Aeronautics and Space Administration, 88th Cong. 1st sess. June 12, 1963; Geiger, "Vandenberg Launch Summary," n.p.

45. "Minuteman Salvo Launch Blasted from Base Silos," *Mesa Missileer*, February 25, 1966, 1; "Second Minuteman Salvo Fired Down Test Range," *Mesa Missileer*, December 23, 1966, 1; Robert D. Bickett, *Space and Missile Test Center Index of Missile Launched on the Western Test Range: 16 December 1958–31 December 1969*, Vandenberg AFB, CA, n.d., 35, 51, SWHO.

46. Burrows, 231, 233; Ruffner, xiv–xv.

47. United States Air Force Fact Sheet, "Air Force PRIME Space Vehicle Launched from Vandenberg AFB," 1966, SWHO.

48. Richard P. Hallion, ed., et al., "From Max Valier to Project PRIME (1924–1967)," Vol. I, in *The Hypersonic Revolution: Eight Case Studies in the History of Hypersonic Technology* (History Office, Aeronautical Systems Division, Wright-Patterson AFB, OH, 1987), v–iii, 633, 695, 714.

49. Ibid., 640.

50. Ibid., 694–95.

51. Ibid., 699.

52. Ibid., 702.

53. Ibid., 711.

54. Ibid., v-iv, 717.

55. Carl Berger, MSgt. Max Gerster, TSgt. Bruce Pollica, and TSgt. Thomas N. Thompson, *History of the 1st Missile Division, 1 January–30 June 1959*, Vandenberg AFB, CA, 1961, 29, 29n, 30, 31, SWHO.

56. "Sergeant's Quick Thinking, Fast Action Save Lives of Airmen in Dire Situation," *SAC Missileer*, February 15, 1963, 3A. Because while oxygen and rocket fuel are highly volatile when they come in contact, liquid nitrogen was substituted for liquid oxygen during training exercises, particularly during rapid dual propellant loading.

57. Ibid.

58. Ibid.; special orders and "Citation to Accompany the Award of Air Force Commendation Medal" sent to the author by Donald J. Sharpless. By the time the newspaper article was published two and a half years after the accident, Kurashewich had been promoted to senior master sergeant and reassigned from the 576th Strategic Mis-

sile Squadron to the 3901st Strategic Missile Evaluation Squadron, also at Vandenberg.

59. Letter with accident summary from Col. Wayne A. Gustafson, 1st Strategic Aerospace Division, to multiple organizations at Vandenberg, August 10, 1961, SWHO; Certificate of Death for Robert Lewis Duncan, August 8, 1961, Clerk Recorder's Office, Santa Barbara County, CA.

60. Carl Berger and SSgt. Ray A. Hanner, Jr., *History of the 1st Missile Division, 1 July–31 December 1960*, Vandenberg AFB, CA, 1961, 76, 78. On April 9, 1959, a Thor missile exploded on the launch pad during a static firing test, and an Atlas exploded during a dual propellant loading on March 5, 1960.

61. SSgt. William Gould, "Liquid Gas in Titan Hole Triggers Dramatic Rescue," *SAC Missileer*, October 21, 1960, 1–2a.

62. Ibid.

63. Ibid.; Berger and Hanner, *History of the 1st Missile Division, 1 July–31 December 1960*, 78.

64. Berger and Hanner, *History of the 1st Missile Division, 1 July–31 December 1960*, 78–79.

65. Ibid., 79. The explosion had settled an ongoing debate that had continued up to the time of the accident about the level of blast door protection needed to protect personnel in Titan silos. Study recommendations had varied widely. The latest findings by the Armour Research Foundation showed the need for stronger protection than was currently in place at the OSTF. Air Force Ballistic Mission Division accepted the recommendation and ordered the 6565th Test Wing at Vandenberg to strengthen the blast lock doors in the OSTF. See pages 86–87.

66. Ibid., 81.

67. Gerald T. Cantwell and Dean J. Stevens, *History of the 1st Strategic Aerospace Division, 1 July–31 December 1964*, Vandenberg AFB, CA, 1965, 62, SWHO; miscellaneous Air Force progress messages pertaining to Titan B-32, SWHO.

68. Cantwell and Stevens, 62–63; miscellaneous Air Force progress messages pertaining to Titan B-32, SWHO.

69. Cantwell and Stevens, 62–63; miscellaneous Air Force progress messages pertaining to Titan B-32, SWHO. A message from the 1st Strategic Aerospace Division's Directorate of Safety to the Air Force Chief of Staff on September 29, 1964, mentioned two additional airmen in the silo, TSgt. Paul T. Seashore and A1C C. L. Snipes. They escaped injury.

70. Cantwell and Stevens, 63.

71. "TF-II DASO Program Interim Report Missile LGM-25C-B32," 6595th Aerospace Test Wing, October 5, 1964, 5, SWHO.

72. Miscellaneous Air Force progress messages pertaining to Titan B-32, SWHO.

73. Cantwell and Stevens, 67.

74. "Report of the 6595th Aerospace Test Wing Missile Accident Investigation Involving Air Force Western Test Range Operation No. 3373 at Van-

denberg AFB," September 24, 1965, SWHO. The pre-launch trajectory data was furnished by the program office to the Air Force Western Test Range organization at Vandenberg, which had assumed the data was valid and was not required to review the information for technical accuracy.

75. Ibid.; letter from Col. H. S. Hensley, Jr., Headquarters Air Force Systems Command (AFSC/SCIZM), to AFSC/SCGR, with two attached letters, January 15, 1966, SWHO. The pre-launch trajectory data was furnished by the program officer to the Air Force Western Test Range organization at Vandenberg, which had assumed the data was valid, and was not required to review the information for technical accuracy.

76. Mark C. Cleary, "Cuban Missile Crisis Anniversary." Part of a collection of 95 pre-published newspaper articles written by Cleary between December 2006 and January 2009 for the base paper at Patrick Air Force Base, Florida. The full collection is titled "This Week in History, 45th Space Wing *Missileer* Newspaper Articles, December 2006–January 2009." Cleary generously sent a copy of this compilation to the author.

77. 6595th Aerospace Test Wing, "Historical Report, 22 October 1962 Through 20 November 1962," Vandenberg AFB, CA, 1962, n.p., SWHO; "Strategic Air Command Increases Alert Posture," *Mesa Missileer*, October 26, 1962, 1; Geiger, "Vandenberg Launch Summary," n.p., SWHO.

78. Peter Carlson, *K Blows Top: A Cold War Comic Interlude Starring Nikita Khrushchev, America's Most Unlikely Tourist* (New York: Public Affairs, 2009), 10, 12.

79. Ibid., 156–60, 168; "Five K-Days," *New York Times*, September 20, 1959, E1; "Premier Annoyed by Ban on a Visit to Disneyland," *New York Times*, September 20, 1959, 20.

80. Carlson, 161–64, 169–71.

81. Ibid., 177–78.

82. Ibid., 180.

83. Ibid.; "Premier, Russ Party Ignore Missile Base," *Los Angeles Times*, September 21, 1959, 2.

84. J. W. Caldwell, *Command History 1961, a Historical Report: U.S. Naval Missile Facility Point Arguello*, U.S. Navy, 1963, 9, SWHO; Richard van Osten, "Vice President Briefed at Ames and Edwards," *Missiles and Rockets*, October 9, 1961, 15.

85. "Attorney General 'Impressed' with Objectives of FCI Here," *Lompoc Record*, March 23, 1962, 1; "Address by Attorney General Robert F. Kennedy, Conference on Crime Prevention in California," U.S. Department of Justice, March 24, 1962, SWHO.

86. Gerald T. Cantwell and Warren S. Howard, *History of the 1st Strategic Aerospace Division, 1 January–30 June 1962*, Vandenberg AFB, CA, 1962, 19–20, SWHO; Harry Crompe, "Base 'Perfect' Host for President's Visit," *Lompoc Record*, March 26, 1962, 1, 2-a.

87. Cantwell and Warren, 19.

88. Ibid., 20; "AFR 80-14 Category III Missile Flight Report 7, Curry Comb I," 1st Strategic Aerospace Division, April 16, 1962, 2-1, II-3, SWHO. Today, Warren AFB is known as F. E. Warren AFB.

89. Cantwell and Howard, 40–43; "AFR 80-14 Category III Missile Flight Report 7, Curry Comb I," 3-1, 3-2.

90. Cantwell and Howard, 20–21; Richard West, "Kennedy Sees Missile Launching, Gets Badge," Los Angeles Times, March 24, 1962, 13.

91. Cantwell and Howard, 21; "AFR 80-14 Category III Missile Flight Report 7, Curry Comb I," 4-1. The other members of the launch and guidance crews were SMSgt. William J. Chupka, SSgt. Roger R. Egert, SSgt. Sherman L. Jackson, SSgt. Reginald E. Bufkin, SSgt. Dale M. Howe, Capt. Kenneth E. Haithcoat, SSgt. Gerald G. Beth, SSgt. Arnold E. Kovhaka, and A1C Franklin D. Waters. These names are listed in the Flight Report. The history report by Cantwell and Howard appears to have misspelled some of the names.

92. Cantwell and Howard, 21–22, 23.

## Epilogue

1. Lompoc Centennial Committee, "Lompoc: The First 100 Years," 1973, 56. The current population is posted on a new city road sign.

2. David Jackson, "Former Soldiers Celebrate Camp Cooke Anniversary," Lompoc Record, October 20, 1991, 1, A10.

3. "Tank Dedication Ceremony Honors Camp Cooke's 50th Anniversary," Space & Missile Times, October 25, 1991, 6.

4. Letter from William A. Hamberg, 5th Armored Division Association, to Col. Daniel R. Clark, 4392nd Aerospace Support Wing, Vandenberg AFB, May 8, 1990, SWHO; letter from Clark to Hamberg, June 14, 1990, SWHO. Both documents in memorial file.

5. Janene Scully, "Remembers Those Activated for the Korean War," Santa Maria Times, September 2, 2000, A-1, A-10; letter from Cassandra M. Brown, Headquarters Air Force Space Command to AFREA/DR, April 3, 2000, SWHO. Both documents in memorial file.

6. Jeffrey Geiger, History of the 30th Space Wing, 1 January 2007 to 31 December 2007, Vandenberg AFB, CA, 2008, 69–73.

7. Geiger, "Vandenberg Launch Summary," n.p., SWHO.

8. Ibid.

9. Ibid.; Jeffrey Geiger, History of the 30th Space Wing, 1 January 2011 to 31 December 2011, Vandenberg AFB, CA, 2012, 125–26. Operational flight testing of the Minuteman III was originally conducted by the Strategic Air Command until 1993 when the program transferred to Air Force Space Command. In 2010 it transferred to Air Force Global Strike Command.

## Appendix A

1. Otis E. Young, The West of Philip St. George Cooke, 1809–1895 (Glendale, CA: Arthur H. Clark, 1955), 21, 23–24.

2. Ibid., 36, 41–42, 44–45, 55–56.

3. Ibid., 63–65, 69, 85–86.

4. Ibid., 104–05, 119–21.

5. Ibid., 179–80; Adrian G. Traas, review of A History May Be Searched in Vain: A Military History of the Mormon Battalion, by Sherman L. Fleek, Army History, Fall 2008.

6. Young, 236–39, 241–42.

7. Ibid., 248–49, 254–55, 265.

8. Ibid., 267–70, 273.

9. Ibid., 294, 300–03, 07, 318–19, 320–21.

10. Ibid., 322–23.

11. Ibid., 330–32, 335–38, 341.

12. Ibid., 342–43, 345.

13. Ibid., 346, 348, 350–53.

14. Ibid., 353, 358.

## Appendix B

1. Carol Anders, "Marshallia Ranch: From Mexican Rancho to Missile Country," Space 'n Lace: Vandenberg Officers' Wives Club Magazine, November 1982, 6.

2. Ibid.

3. Ibid., 6–7; Averam B. Bender, "From Tanks to Missiles: Camp Cooke/Cooke Air Force Base (California), 1941–1958," Arizona and the West, Autumn 1967, 221.

4. First Strategic Aerospace Division History Office (1 STRAD/HO), "The Story of Marshallia Ranch: A History Dating Back to the Olivera Adobe Built in 1837," Vandenberg AFB, CA, n.d., 1.

5. Ibid., 2.

6. Ibid.

7. Ibid., 2–3.

8. Ibid., 3.

9. Ibid.

10. Ibid., 3.

11. Ibid., 3–4.

12. Ibid., 4; Anders, 7.

13. STRAD/HO, 4; Eve Allgair, "Story of a Ranch," Santa Barbara News-Press, August 16, 1953, G-8-9.

14. Allgair, G-9.

15. STRAD/HO, 4.

16. Allgair, G-9; "Marshallia Ranch to Reopen on July 4," Santa Maria Daily Times, June 27, 1947, 1.

17. Allgair, G-9; "Marshallia Ranch to Reopen on July 4," June 27, 1947, 1.

18. Anders, 8.

## Appendix D

1. The 50th Armored Infantry Regiment of the 6th Armored Division arrived at Camp Cooke on March 19, 1943. It was redesignated the 50th

Armored Infantry Battalion, less the 1st and 2nd Battalions, on September 20, 1943. The 1st and 2nd Battalions were redesignated the 44th and 9th Battalions, respectively. See Stanton, *Order of Battle*, 209.

2. Arrived at Camp Cooke on March 20, 1943,

as the 69th Armored Regiment and was redesignated (less the 3rd Battalion, Maintenance Company, Service Company, and Reconnaissance Squadron) as the 69th Tank Battalion on September 20, 1943. See Stanton, *Order of Battle*, 291.

# BIBLIOGRAPHY

## Articles

Anders, Carol. "Marshallia Ranch: From Mexican Rancho to Missile Country." *Space 'n Lace: Vandenberg Officers' Wives Club Magazine* (November 1982): 6–8.

Bender, Averam B. "From Tanks to Missiles: Camp Cooke/Cooke Air Force Base (California), 1941–1958." *Arizona and the West* (Autumn 1967): 219–42.

National Park Service. "Radar Station B-71 Trinidad Radar Station California." In Cohen, *V for Victory*, 208.

## Newspapers and Magazines

*The Billboard*
*Cooke Clarion* (alternately the *Camp Cooke Clarion* and *The Clarion*)
*Daily Variety*
*Down Beat*
*Hollywood Reporter*
*Lompoc Record*
*Los Angeles Daily News*
*Los Angeles Times*
*Missiles and Rockets*
*National Enquirer*
*New York Times*
*SAC Missileer* (Renamed *Mesa Missileer* and later *Space & Missile Times*)
*Santa Barbara News-Press*
*Santa Maria Daily Times* (Renamed *Santa Maria Times* after 1948)
*Theater Arts*
*Yank: The Army Weekly*

## Booklets and Pamphlets

Armed Forces Advertising Service. *This Is ... Camp Cooke, California*, 1951.

Army Special Services. *Camp Cooke Welfare Carnival*, October 1952.

*A Camera Trip Through Camp Cooke: A Pictorial Book of the Camp and Its Activities.* Brooklyn, NY: Ullman, 1943.

First Strategic Aerospace Division History Office. "The Story of Marshallia Ranch: A History Dating Back to the Olivera Adobe Built in 1837." Historical vignette, Vandenberg AFB, CA, n.d.

Headquarters Signal Office, Camp Cooke. *Highlights of Camp Cooke U.S. Army California*, circa 1944.

Hollywood Victory Committee. *HVC*, circa 1945.

Kelly, Sgt. Bob, ed. 11th Armored Division Special Service Office. *Meet Camp Cooke*, 1944.

Keough, Lt. Emmett L., Sgt. Bob Kelly, Sgt. Noel A. Vicentini, Pfc. Robert Washington Race, Pfc Anthony Stanziola, and T/5 John P. Thomas. *Eleventh Armored Division.* Hollywood, CA: Murray & Gee, 1944.

Peebles, Curtis. *High Frontier: The United States Air Force and the Military Space Program.* U.S. Air Force History and Museum Program, 1997.

*6th Armored Division: A Pictorial History of the Super 6th, February 15, 1942* [to] *February 15, 1944.* n.p., circa 1944.

## Books

Alicoate, Jack, ed. *The 1942 Film Daily Year Book of Motion Pictures, 24th Annual Edition.* Fort Lee, NJ: J. E. Brulatour, 1943.

_____, ed. *The 1943 Film Daily Year Book of Motion Pictures, 25th Annual Edition.* Fort Lee, NJ: J. E. Brulatour, 1944.

_____, ed. *The 1944 Film Daily Year Book of Motion Pictures, 26th Annual Edition.* Fort Lee, NJ: J. E. Brulatour, 1945.

Andrews, Maxene, and Bill Gilbert. *Over Here, Over There: The Andrews Sisters and the USO Stars in World War II.* New York: Zenith Books, 1993.

Appleman, Roy E. *South to the Naktong, North to the Yalu (June–November 1950).* Washington, D.C.: Center of Military History, U.S. Army, 1961.

Association of Air Force Missileers, with David K. Stumpf. *Air Force Missileers.* Paducah, KY: Turner Publishing, 1998.

Bak, Richard. *Joe Louis: The Great Black Hope.* Dallas, TX: Taylor Publishing, 1996.

Burrows, William E. *Deep Black: Space Espionage and National Security.* New York: Random House, 1986.

Cameron, Robert Steward. *Mobility, Shock, and Firepower: The Emergency of the U.S. Army's Armor Branch, 1917–1945.* Washington, D.C.: Center of Military History, U.S. Army, 2008.

Carlson, Peter. *K Blows Top: A Cold War Comic Interlude Starring Nikita Khrushchev, America's Most Unlikely Tourist.* New York: Public Affairs, 2009.

Cline, Ray S. *Washington Command Post: The Operations Division.* Washington, D.C.: Center of Military History, U.S. Army, 2003.

Cohen, Stan. *V for Victory: America's Home Front During World War II.* Missoula, MT: Pictorial Histories Publishing, 1991.

Craven, Wesley F., and James L. Cate, eds. *The Army Air Forces in World War II*, 7 vols. Chicago: University of Chicago Press, 1948–1958. Reprint, Washington, D.C.: Office of Air Force History, 1983. See Vol. 1, *Plans and Early Operations, January 1939 to August 1942*, and Vol. 7, *Service Around the World.*

Cullen, Frank, with Florence Hackman and Donald McNeilly. *Vaudeville, Old and New: An Encyclopedia of Variety Performers in America*, 2 vols. New York: Routledge, 2007.

DeNevi, Donald. *The West Coast Goes to War, 1941–1942.* Missoula, MT: Pictorial Histories Publishing, 1998.

Dunning, John. *On the Air: The Encyclopedia of Old-Time Radio.* New York: Oxford University Press, 1998.

_____. *Tune in Yesterday: The Ultimate Encyclopedia of Old-Time Radio, 1925–1976.* Englewood Cliffs, NJ: Prentice Hall, 1976.

Epstein, Daniel Mark. *Sister Aimee: The Life of Aimee Semple McPherson.* New York: Harcourt Brace Jovanovich, 1993.

Fine, Lenore, and Jesse A. Remington. *The Corps of Engineers: Construction in the United States.* Washington, D.C.: Center of Military History, U.S. Army, 1972.

*44th Infantry Division, Camp Cooke, California.* Baton Rouge, LA: Army and Navy Publishing, circa 1953.

Gilbert, Adrian. *The Encyclopedia of Warfare: From Earliest Times to the Present Day.* Guilford, CT: Lyons Press, 2003.

Greenfield, Kent Roberts, Robert R. Palmer, and Bell I. Wiley. *The Organization of Ground Combat Troops.* Washington, D.C.: Center of Military History, U.S. Army, 1987.

Hanson, R. L. *The Boys of Fifty: The 625th Field Artillery Battalion, 40th Infantry Division California National Guard 1946–1954.* Raleigh, NC: LuLu Press, 2006.
Hartt, Julian. *The Mighty Thor: Missile in Readiness.* New York: Duell, Sloan and Pearce, 1961.
Hermes, Walter G. *Truce Tent and Fighting Front.* Washington, D.C.: Center of Military History, U.S. Army, 1966.
Hewes, James E. Jr. *From Root to McNamara: Army Organization and Administration, 1900–1963.* Washington, D.C.: Center of Military History, U.S. Army, 1975.
Kellner, Bruce. *The Harlem Renaissance.* Westport, CT: Greenwood Press, 1984.
Mauer, Mauer, ed. *Air Force Combat Units of World War II.* Washington, D.C.: Office of Air Force History, 1983.
Mayo, Lida. *The Ordnance Department: On Beachhead and Battlefront.* 1968. Reprint, Washington, D.C.: Center of Military History, U.S. Army, 1991.
McReynolds, John V. *Vanished: Lompoc's Japanese.* Lompoc, CA: Press Box Productions, 2010.
Millett, John D. *The Organization and Role of The Army Service Forces.* Washington, D.C.: Center of Military History, U.S. Army, 1987.
NASA Historical Staff. *Astronautics and Aeronautics, 1963: Chronology on Science, Technology, and Policy.* Washington, D.C.: National Aeronautics and Space Administration, 1964.
Neufeld, Jacob. *The Development of Ballistic Missiles in the United States Air Force, 1945–1960.* Washington, D.C.: Office of Air Force History, U.S. Air Force, 1990.
Ordway III, Frederick I., and Mitchell R. Sharpe. *The Rocket Team.* New York: Thomas R. Cromwell, 1979.
Palmer, Robert R., Wiley I. Bell, and William R. Keast. *Procurement and Training of Ground Combat Troops.* Washington, D.C.: Center of Military History, U.S. Army, 1991.
Preston, Jack, and Mrs. Edward G. Robinson. *The Desert Battalion.* Hollywood, CA: Murray & Gee, 1944.
Sawicki, James A. *Antiaircraft Artillery Battalions of the U.S. Army,* 2 vols. Dumfries, VA: Wyvern Publications, 1991.
_____. *Field Artillery Battalions of the U.S. Army,* 2 vols. Dumfries, VA: Centaur Publications, 1978.
_____. *Infantry Regiments of the U.S. Army.* Dumfries, VA: Wyvern Publications, 1981.
_____. *Tank Battalions of the U.S. Army.* Dumfries, VA: Wyvern Publications, 1983.
*747th Amphibious Tank and Tractor Battalion, Camp Cooke, California.* Baton Rouge, LA: Army and Navy Publishing, circa 1952.
Sies, Luther F. *Encyclopedia of American Radio, 1920–1960.* Jefferson, NC: McFarland, 2008.
Stanton, Shelby L. *Order of Battle U.S. Army, World War II.* Novato, CA: Presidio Press, 1984.
Starr, Kevin. *Embattled Dreams: California in War and Peace 1940–1950.* New York: Oxford University Press, 2002.
Terrett, Dulany. *The Signal Corps: The Emergency.* Washington, D.C.: Center of Military History, U.S. Army, 1956.
Tucker, Sherrie. *Swing Shift: "All Girl" Bands of the 1940s.* Durham, NC: Duke University Press, 2000.
U.S. Army, 40th Infantry Division Public Information Office. *Fortieth in Review.* Tokyo, Japan, 1952.
Villahermosa, Gilberto N. *Honor and Fidelity: The 65th Infantry in Korea, 1950–1953.* Washington, D.C.: Center of Military History, U.S. Army, 2009.
Wagner, Margaret E., Linda B. Osborne, Susan Reyburn, and the Staff of the Library of Congress, eds. *World War II Companion.* New York: Simon & Schuster, 2007.
Watson, Robert J. "History of the Office of the Secretary of Defense." In *Into The Missile Age, 1956–1960, Vol. IV,* edited by Alfred Goldberg. Washington, D.C.: Historical Office, Office of the Secretary of Defense, 1997.
Wertheim, Arthur Frank. *Radio Comedy.* New York: Oxford University Press, 1979.
Young, Otis E. *The West of Philip St. George Cooke, 1809–1895.* Glendale, CA: Arthur H. Clark, 1955.

## *Congressional Documents*

Private Relief Bill for Vertie Bea Loggins, 1945–1946, Folder 3, Box 101, Helen Gahagan Douglas Collection, Carl Albert Congressional Research and Studies Center, University of Oklahoma.

U.S. House of Representatives. Committee on Claims. *Vertie Bea Loggins.* 79th Cong., 2nd sess., 1946. H. Rep. 1762.
U.S. House of Representatives. Committee on Science and Astronautics. *Astronautical and Aero- nautical Events of 1962.* Report prepared by the National Aeronautics and Space Adminis- tration. 88th Cong. 1st sess. June 12, 1963.
U.S. Senate. Committee on Aeronautical and Space Sciences. *Manned Orbiting Laboratory.* 89th Cong., 2nd sess. February 24, 1966.
U.S. Senate. Committee on Claims. *Vertie Bea Loggins.* 79th Cong., 2nd sess., 1946. S. Rep. 1290.

## *Official U.S. Air Force and U.S. Navy Unit Histories*

Berger, Carl. *History of the 1st Missile Division.* Vandenberg AFB, CA, n.d.
Berger, Carl, MSgt. Max Gerster, TSgt. Bruce Pollica, and TSgt. Thomas N. Thompson. *History of the 1st Missile Division, 1 January–30 June 1959.* Vandenberg AFB, CA, 1961.
_____. *History of the 1st Missile Division, 1 July–31 December 1959.* Vandenberg AFB, CA, 1960.
Berger, Carl, and SSgt. Ray A. Hanner, Jr. *History of the 1st Missile Division, 1 July–31 December 1960.* Vandenberg AFB, CA, 1961.
Berger, Carl, Warren S. Howard, and SSgt. Ray A. Hanner, Jr. *History of the 1st Missile Division, 1 January–30 June 1961.* Vandenberg AFB, CA, 1961.
Berger, Carl, and TSgt. Bruce Pollica. *History of the 1st Missile Division, 1 January–30 June 1960.* Vandenberg AFB, CA, 1960.
Bickett, Robert D. *History of the Air Force Western Test Range, 15 May–31 December 1964.* Van- denberg AFB, CA, n.d.
_____. *History of the Air Force Western Test Range, 1 January–31 December 1966.* Vandenberg AFB, CA, n.d.
_____. *History of the Air Force Western Test Range, 1 January–30 June 1969.* Vandenberg AFB, CA, n.d.
_____. *Space and Missile Test Center Index of Missiles Launched on the Western Test Range, 16 December 1958–31 December 1969.* Vandenberg AFB, CA, n.d.
Caldwell, J. W. *Command History 1960, a Historical Report: U.S. Naval Missile Facility Point Arguello.* U.S. Navy, 1962.
_____. *Command History 1961, a Historical Report: U.S. Naval Missile Facility Point Arguello.* U.S. Navy, 1963.
Cantwell, Gerald T., and Warren S. Howard. *History of the 1st Strategic Aerospace Division, 1 Jan- uary–30 June 1962.* Vandenberg AFB, CA, 1962.
_____, and Dean J. Stevens. *History of the 1st Strategic Aerospace Division, 1 July–31 December 1964.* Vandenberg AFB, CA, 1965.
Cleary, Mark C. *The 6555th: Missile and Space Launches Through 1970.* 45th Space Wing History Office, Patrick AFB, FL, November 1991.
Donahue, Joseph T. Jr., and Dean J. Stevens. *History of the 1st Strategic Aerospace Division, Janu- ary–June 1966.* Vandenberg AFB, CA, n.d.
Gerster, MSgt. Max. *History of the 1st Missile Division and the 392nd Air Base Group, 15 April 1957 through 30 June 1957.* Cooke AFB, CA, 1957.
_____. *History of the 1st Missile Division, 1 July through 31 December 1957.* Cooke AFB, CA, 1958.
Geiger, Jeffrey. *History of the 30th Space Wing, 1 January 2007 to 31 December 2007.* Vandenberg AFB, CA, 2008.
_____. *History of the 30th Space Wing, 1 January 2011 to 31 December 2011.* Vandenberg AFB, CA, 2012.
"Headquarters, 3265th Training Squadron." United States Air Force Historical Research Agency, Maxwell Air Force Base, Montgomery, Alabama. Microfilm roll K0707, index 1944.
Historical Division. *History of Air Research & Development Command, 1 July–31 December 1960, Vol. III.* Andrews AFB, MD, 1961.
Historical Report. "6565th Test Wing (Development), 20 October 1960 through 31 December 1960." Vandenberg AFB, CA, n.d.
Menken, Arthur. *History of the Pacific Missile: 1 November 1945 to 30 June 1959.* U.S. Navy, Point Mugu, CA, n.d.

6595th Aerospace Test Wing. *Historical Report, 22 October 1962 through 20 November 1962.* Vandenberg AFB, CA, 1962.

Space Division History Office. *Space and Missile Systems Organization: A Chronology, 1954–1979.* Los Angeles AFB, CA, n.d.

## National Archives Records

Material about Camp Cooke in the National Archives and Records Administration at the College Park, Maryland, branch is sparse and scattered across several Record Groups (RG). For this work, I consulted RG 77 (Chief of Engineers), RG 112 (Office of the Surgeon General), RG 337 (HQ Army Ground Forces), RG 338 (U.S. Army Operational, Tactical, and Support Organizations—World War II and Thereafter), and RG 389 (Office of the Provost Marshal General).

## Unpublished Reports

Hollywood Coordinating Committee, 1943–1953, Association of Motion Picture and Television Producers (AMPTP) records.

Hollywood Coordinating Committee talent records, Association of Motion Picture and Television Producers (AMPTP) records.

Hollywood Victory Committee monthly reports, June 1942–December 1945, MPAA Breen Office files.

Margaret Herrick Library, Academy of Motion Picture Arts and Sciences, Beverly Hills, CA.

United Service Organizations (USO) Camp Shows, 1941–1947, Association of Motion Picture and Television Producers (AMPTP) records.

United Service Organizations, Camp Shows Publicity Records, 1941–1957, Hollywood Coordinating Committee records.

United Service Organizations (USO) Camp Shows, 1951–1954, Association of Motion Picture and Television Producers (AMPTP) records.

## Miscellaneous Manuscripts and Studies

Army Post Engineer. "Camp Cooke Building Index." May 15, 1955, author's collection.

Barrett, M., J. Goetz, J. Klemm, E. Ortlieb, and C. Thomas. *Analysis of Radiation Exposure for Military Participants: Exercises Desert Rock 1, II, and III—Operation Buster-Jangle.* Sciences Applications International Corp, McLean, Virginia, December 1987.

Bob Hope script, March 3, 1942. Milton Josefberg papers, collection 3154, box 7, American Heritage Center, University of Wyoming.

Fink, Ron. "The Ocean Park—Surf Beach Docent Handbook." Ocean Beach Commission, Santa Barbara County, 2001.

Gebhard, David, and David Bricker. "The Former U.S. Coast Guard Lifeboat Rescue Station and Lookout Tower Point Arguello, California: (1936–1941), a Public Report." University of California, for Heritage Conservation and Recreation Service Interagency Archaeological Services, n.p., 1980.

Hallion, Richard P., ed., with John Becker, Ronald Boston, Clarence Geiger, Richard Hallion, Robert Houston, and John Vitelli. "The Hypersonic Revolution: Eight Case Studies in the History of Hypersonic Technology." History Office, Aeronautical Systems Division, Wright-Patterson AFB, Ohio, 1987. See Vol. 1, "From Max Valier to Project PRIME (1924–1967)."

Heberling, Lynn O'Neal. "Soldiers in Greasepaint: USO–Camp Shows, Inc., During World War II." Ph.D. diss., Kent State University, 1989.

Kahn, Ava F. "Historic Background of Red Roof Canyon." Social Process Research Institute, Public Archaeology, University of California, Santa Barbara, August 1981.

Lee Keeling and Associates, Inc. "First Gross Mineral Appraisal [of] Vandenberg Air Force Base, Santa Barbara County, California." Tulsa, OK, 1977.

_____. "Second Gross Mineral Appraisal, [of] Vandenberg Air Force Base. Santa Barbara County, California." Tulsa, OK, 1977.

Lompoc Centennial Committee. "Lompoc: The First 100 Years." Lompoc Library, 1973.

Ponton, Jean, Stephen Rohrer, Carl Maag, and Jean Massie. *Shots, Sugar and Uncle: The Final Tests of the Buster-Jangle Series 19 November–29 November 1951.* JRB Associates, McLean, VA, June 23, 1982.

    Purkiss, Wesley W., ed. "A History of Camp Cooke, 1941 to 1946." Vandenberg AFB Library, April 1946.

"Reusable Launch Vehicle/Spacecraft Concept." Report No. GDC-DCB-68–017, General Dynamics Convair Division, November 1968.

Ruffner, Kevin C., ed. *Corona: America's First Satellite Program.* Washington, D.C.: Central Intelligence Agency, 1995.

U.S. Army. *Station List of the Army of the United States, June 1, 1943.* California State Military Museum.

## Other Documents

Amendment No. 1 to Burke/LeMay agreement, December 31, 1960. History Office, 30th Space Wing, Vandenberg AFB, CA.

Anonymous background paper. "Background of Vandenberg/Arguello Relationships," n.d., History Office, 30th Space Wing, Vandenberg AFB, CA. History Office, 30th Space Wing, Vandenberg AFB, CA.

Cleary, Mark C. "Cuban Missile Crisis Anniversary." Part of a collection of 95 pre-published newspaper articles written by Cleary for the base paper at Patrick Air Force Base, Florida. Cleary sent the author the full collection of articles titled "This Week in History, 45th Space Wing *Missileer* Newspaper Articles, December 2006–January 2009."

Death Certificate for Robert Lewis Duncan, August 8, 1961. Clerk Recorder's Office, Santa Barbara County, CA.

Geiger, Jeffrey. "Vandenberg Launch Summary," n.d. History Office, 30th Space Wing, Vandenberg AFB, CA.

Monroe, Admiral Jack P. "Pacific Missile Range." Presentation to Walker L. Cisler, October 1959. History Office, 30th Space Wing, Vandenberg AFB, CA.

## Internet Sites

Allan Hancock College. www.hancockcollege.edu
American Film Institute Catalog of Feature Films. www.afi.com
California State Military Museum. www.militarymuseum.org
Circus Historical Society. www.circushistory.org
Congressional Medal of Honor Society. www.cmohs.org
Defense Nuclear Agency. www.dtra.mil
Dignity Memorial. www.dignitymemorial.com
The 11th Armored Division Legacy Group. www.11tharmoreddivision.com
Hillbilly Music. www.hillbilly-music.com
Illinois National Guard. www.il.ngb.army.mil
Internet Archive. http://archive.org
Internet Movie Database. www.imbd.com
Korean War Educator. www.koreanwar-educator.org
Lambiek Comiclopedia. http://lambiek.net
National Guard Education Foundation. www.ngef.org
Super Sixth: The 6th Armored Division in World War II. www.super6th.org
Turner Classic Movies. www.tcm.com
The 20th Armored Division Association. www.20tharmoreddivision.net
United Service Organizations. www.uso.org
U.S. Army Center of Military History. www.history.army.mil
U.S. Defense Threat Reduction Agency. www.dtra.mil
U.S. Department of Defense. www.dod.mil
U.S. Department of State, Office of the Historian. http://history.state.gov
Vandenberg Air Force Base. www.vandenberg.af.mil
Wikipedia. www.wikipedia.org

# INDEX

Numbers in **bold italics** refer to pages with photographs.